# Energy Efficiency: Perspectives on Individual Behavior

**American Council for an Energy-Efficient Economy Series on Energy Conservation and Energy Policy**

**Series Editor: Carl Blumstein, Universitywide Energy Research Group, University of California**

*Financing Energy Conservation*
*Energy Efficiency in Buildings: Progress and Promise*
*Energy Efficiency: Perspectives on Individual Behavior*

# Energy Efficiency: Perspectives on Individual Behavior

Editors
Willett Kempton
Max Neiman

American Council for an Energy-Efficient Economy
Washington, DC and Berkeley, CA

in cooperation with:
Universitywide Energy Research Group
University of California
Family Energy Project, Institute for Family and Child Studies
Michigan State University

1987

Cover art and design by Allison Turner. Book Design by Paula Morrison. All figures by A Small Creative Agency.

Portions of this book have been reprinted with the permission of other publishers. "Why Don't People Weatherize Their Homes?: An Ethnographic Solution," by Richard Wilk and Harold Wilhite, is reprinted from *Energy* 10(5):621-629, 1985, with permission of Pergamon Press Ltd. "Optimizing Investment Levels for Energy Conservation: Individual Versus Social Perspective, and the Role of Uncertainty, by Ari Rabl is reprinted from *Energy Economics*, January 1985, pp.259-264, with permission of Butterworth Scientific. Portions of "Determinants of Household Energy Efficiency and Demand," by Leslie Baxter et al., were drawn from an article entitled "Efficiency Analysis of Household Energy Use" that appeared in the January 1986 issue of *Energy Economics*. "Residential Hot Water: A Behaviorally Driven System," by Willett Kempton is an expanded version of a paper also appearing in *Energy: The International Journal.*

**Library of Congress Cataloguing in Publication Data**

Energy Efficiency: Perspectives on Individual Behavior

(Series on energy conservation and energy policy)
Includes bibliography and index.
1. Energy consumption—United States. 2. Energy
Conservation—United States. 3. Consumers—United States.
4. Energy Policy—United States. I. Kempton, Willett,
1984- . II. Neiman, Max. III. Series.
HD9502.U52E4883 1987 333.79'16'0973 86-28817
ISBN 0-918249-05-8

First published in 1987

American Council for an Energy-Efficient Economy
1001 Connecticut Ave., NW, Suite 535, Washington, DC 20036

Printed in the United States of America

10 9 8 7 6 5 4 3 2 1

## TABLE OF CONTENTS

# PREFACE

This book stems from the 1984 Summer Study on Energy Efficiency in Buildings, sponsored by the American Council for an Energy-Efficient Economy (ACEEE) and held in Santa Cruz, California. At the Summer Study, this volume's authors presented papers and participated in panel discussions on behavioral aspects of energy conservation. Because of the strong interest in the behavioral studies at Santa Cruz, ACEEE asked us to select papers to be revised and included in a volume intended to provide broad treatment of experience in this evolving field.

The editors would like to acknowledge the support and encouragement of Carl Blumstein, who served as chairman of the Summer Study and who originally suggested the volume. Our appreciation is also extended to Jon Koomey and Peter DuPont for their contributions as copy editor and production manager (respectively).

The editors also acknowledge the financial support that made this volume possible—from the Universitywide Energy Research Group (of the University of California) to Max Neiman, and from the Michigan State University Family Energy Project to Willett Kempton.

Finally, we are grateful to the many organizations that supported the 1984 summmer study: Bonneville Power Administration, Electric Power Research Institute, Gas Research Institute, Lawrence Berkeley Laboratory, Michigan State University, Oak Ridge National Laboratory, Pacific Gas & Electric Company, Southern California Edison Company, Standard Oil Company (Ohio), U.S. Department of Energy, University of California, Western Area Power Administration (all sponsors); California Energy Commission and North Carolina Alternative Energy Corporation (both major contibutors); and Owens-Corning Fiberglass Corporation, Governor's Energy Council of Pennsylvania, and Time Energy Systems, Inc. (all contributors).

*Willett Kempton*
*Max Neiman*

## CONTRIBUTING AUTHORS

**Dane Archer**
Department of Sociology
University of California
Santa Cruz, CA

**Lester Baxter**
California Energy Commission
Sacramento, CA

**Barbara Burt**
Arizona State University, West
Tempe, AZ

**Mark Costanza**
University of California
Santa Cruz, CA

**Richard Diamond**
Lawrence Berkeley Laboratory
Berkeley, CA

**Rita Erickson**
University of Minnesota
St. Paul, MN

**Mary Fagerson**
Minnesota Department of Energy
St. Paul, MN

**Shel Feldman**
National Analysts
Philadelphia, PA

**Stephen Feldman**
University of Pennsylvania
Philadelphia, PA

**Bruce Hackett**
Department of Sociology
University of California
Davis, CA

**Dick Holt**
U.S. Department of Energy
Washington, DC

**Bonita Iritani**
University of California
Santa Cruz, CA

**Willett Kempton**
Center for Energy and
Environmental Studies
Princeton University
Princeton, NJ

**Shirlee Krabacher**
Michigan State University
East Lansing, MI

**Joseph Laquatra**
New York College of
Human Ecology
Cornell University
Ithaca, NY

**Mark Levine**
Lawrence Berkeley Laboratory
Berkeley, CA

**James McMahon**
Lawrence Berkeley Laboratory
Berkeley, CA

**Bonnie Morrison**
Institute for Child and
Family Study
Michigan State University
East Lansing, MI

**Max Neiman**
Department of Political Science
University of California
Riverside, CA

**Thomas Pettigrew**
University of California
Santa Cruz, CA

**Ari Rabl**
Center for Energy and
Environmental Studies
Princeton University
Princeton, NJ

**Henry Ruderman**
Lawrence Berkeley Laboratory
Berkeley, CA

**Arie Schinnar**
University of Pennsylvania
Philadelphia, PA

**Ed Vine**
Lawrence Berkeley Laboratory
Berkeley, CA

**Iain Walker**
University of California
Santa Cruz, CA

**Jeff Weihl**
Institute for Child and
Family Study
Michigan State University
East Lansing, MI

**Lawrence White**
University of California
Santa Cruz, CA

**Harold Wilhite**
Resource Policy Group
Oslo, Norway

**Richard Wilk**
Department of Anthropology
New Mexico State University
Las Cruces, NM

**Robert Wirtshafter**
University of Pennsylvania
Philadelphia, PA

## ABOUT THE EDITORS

*Willett Kempton is a cognitive anthropologist with an interest in technology and policy issues. He is now at the Center for Energy and Environmental Studies at Princeton University. Working with a group of engineers and policy analysts, he studies the cognitive and behavioral aspects of energy consumption. Recent work includes mental models of thermostats and home heat loss, everyday arithmetic used to compute changes in energy bills, and heating and cooling behavior in multi-family buildings. Outside of the energy area, he has studied folk classification by indigenous Mexican groups and novice comprehension of computer programs. He received a Ph.D. in cognitive anthropology from the University of Texas at Austin. He has previously held postdoctoral or research faculty positions at the University of California, Berkeley and Michigan State University.*

*Max Neiman is an associate professor of political science at the University of California, Riverside. His research focuses on political factors affecting community and citizen receptivity to energy policy innovation and resource conservation programs. He is currently working on a synthesis of his research into a monograph on energy politics in California from the early 1970s through approximately 1981. His other research interests include urban politics and policy, as well as economic applications to institutional design. He received his Ph.D. in political science from the University of Wisconsin, Milwaukee in 1973.*

# SECTION I:

# Introduction

# SECTION I: Introduction

*Max Neiman*
*University of California, Riverside*

A little more than a decade ago, when energy use and energy conservation first became important topics, many analysts believed that energy consumption could be easily explained using physical data such as climatic conditions and appliance efficiencies. However, these analysts soon discovered large variations in energy use that were not fully explained by the physical data. Identical houses often had substantially different utility bills, and statistical analyses based on physical attributes consistently yielded a large "unexplained" variance. We now ascribe this unexplained variance to differences in human behavior.

Analyzing and explaining human behavior is a difficult but rewarding task. The chapters in this book, written by leading researchers, program analysts, and policy makers, present a variety of perspectives on such analysis. No one perspective explains all behavior, but each is useful in attacking certain problems. From detailed observation of thermostat settings over time to cross-cultural comparisons of energy use, the authors explore models of energy behavior that provide valuable assistance in planning more effective programs and researching their effects.

This book has something to interest almost everyone involved in energy and behavioral sciences. Researchers will find the theoretical analysis fascinating. Social scientists will find insights into human behavior that are transferable to other situations. Program managers and policy makers will find the analysis of observed behavior useful when creating future conservation programs. All readers will be intrigued by the diversity of approaches and ideas. In short, the chapters in this volume represent important contributions to our understanding of energy consumption and human behavior, and are fascinating reading as well.

## ORGANIZATION OF THIS VOLUME

This book is divided into four sections. The first, "Consumer Investment Decision Processes", focuses on the conditions under which individuals invest in energy-efficiency improvements. Some of the works primarily address economic models of behavior, although they clearly do not agree on results. The degree to which individuals calculate the relative costs and benefits of energy-efficiency improvements and whether the putative benefits of enhanced energy efficiency are reflected in the selling prices of households are among the issues addressed in this section. Another group of articles in the section concentrates on social, psychological, and lifestyle models of behavior in an attempt to assess the circumstances under which individuals invest in energy efficiency.

The second section, "The Inference of Individual Behavior from Aggregrate Data", takes several approaches. In some cases the focus is on estimating energy-related variables. Other concerns include estimating the value of energy-efficiency improvements among households, generalizing about thermostat management from differing studies on the subject, and judging the political feasibility of policies to promote solar energy. This heterogeneous section highlights a number of challenges and possibilities associated with large-scale studies of energy behavior, as well as the difficulties of deriving generalizations about individual energy behavior from disparate research approaches.

The third section, "Home Management", includes some genuinely original data and approaches to the study of energy behavior, including ethnographic observation and the use of extensive monitoring of individual households. The chapters examine the great variety of factors that differentiate households in terms of their energy use, including the work schedules of household members and the social functions of various energy uses.

The final section, "The Interaction of Building Systems with

Occupants", also sheds light on conventional wisdom and suggests some limits as well as possibilities regarding public policy. On the one hand, shifting from master to individual metering of energy use appears quite promising as a method of reducing energy demand. On the other hand, the causal view that building design will lead to major changes in energy use, while certainly not disputed here, is modified to include the influence of individual household life-style on energy behavior. These studies suggest that the level of energy savings due to improved design and construction will be affected by the life-style variations among households.

## ALTERNATIVE MODELS OF BEHAVIOR

The chapters in this book offer contrasting conceptions of the motivations behind energy behavior. On the one hand, much emphasis has been placed on attitudinal models of behavioral change. Attitudinal models suggest that a change of individual attitudes must precede behavioral change. These models suggest that education and information programs are needed to alter people's perception of energy use and energy sources and to induce them to change their behavior. Often such efforts involve the notion that unnecessary energy use is "fuelish" or further exhortations designed to make intense energy use seem less fashionable and less linked to progress or modernity. The use of energy fairs and demonstration projects to provide information on state-of-the-art, unconventional energy use are examples of actions designed to influence behavior in part through altering attitudes.

Related to the attitudinal model is the class of research using an economic or rational model of behavior. In this case, changes in energy-relevant behavior are assumed to flow from changes in market conditions. This type of model implies that the consumer is primarily interested in personal utility maximization, and further implies that energy behavior can most effectively be changed by using economic incentives. A wide variety of public policies are premised on the notion that people respond to such incentives. Low-interest loans and tax breaks for alternative energy systems and conservation improvements are designed to make certain energy-related behaviors more attractive.

In some respects, energy outreach programs combine both attitudinal and rational models of behavior; such programs are based on research indicating that fostering energy-efficient behavior is most successfully accomplished through personal contact. Such personalized outreach efforts as the Residential Conservation Service (RCS) appear to be more effective than information alone in assuring that energy consumers have the knowledge to adopt cost-effective energy conservation measures. Information, economic incentives, and attitudes are presumed to interact to produce individual-level changes in energy behavior, as well as structural changes in the market for energy efficiency in homes.

## ISSUES RAISED IN THIS VOLUME

Several research issues are addressed by this volume. First is the problem of the scale of research. There is always a tension between the desire to generalize from statistically sound data using large samples, and the desire to examine and monitor as many aspects as possible of actual energy-related behaviors (e.g., heating and cooling preferences, thermostat adjustments, and water use in cleaning and personal cleanliness). Several of the works in this volume illustrate this tension. Baxter et al, for example, in their presentation of a method for assessing household energy use, employ the Household Screener Survey, involving thousands of households.

On the other hand, Erickson, in her investigation of the cultural determinants of energy use, analyzes a total of 37 households from both the American and Swedish communities. The chapters by Kempton and by Kempton and Krabacher examine seven households in unprecedented detail, exploring these households' hot water use and thermostat management. The Weihl chapter reports the finding concerning energy use and the schedules of household members from five households.

Of course, the issue of generalizability is important when small samples are used to produce highly detailed observations of actual behavior. Perhaps we expect too much from single studies. We often act as if the substantive issues must be resolved in the context of a single study. Consequently, one strives to design a project with large enough samples and with ideal measures of key variables, both explanatory and potentially confounding, so that generalizing across time and place is warranted.

But what ever happened to the idea of replication? It is overly optimistic to hope for a single research team, with sufficient skill and status, to garner a large enough grant to design the single study that produces complete information on energy behavior. Instead it will be the multiple efforts of many scholars, replicating work in different settings, that will more effectively produce knowledge. We cannot resolve the issue here, but clearly the consumers of energy behavior studies must make a place not only for the dramatic, large-scale research expedition. We must also accomodate the smaller-scale studies, which can contribute to our understanding in the long term.

The chapters in this volume raise other important issues. Burt and Neiman, in their analysis of local energy regulation to promote solar energy, raise questions of political feasibility in translating the rhetoric of local roles in energy policy into specific policy actions. Laquatra's analysis argues that rational, economically motivated individuals will adopt certain energy investments (e.g., a more efficient heating system), assuming the market or public policy produce adequate returns on these investments. Also using a rational model perspective, Holt examines how energy efficiency is capitalized into the resale value of homes, while Rabl assesses the problems of uncertainty for consumer rationality. Ruderman, Levine, and McMahon assess market imperfections in producing optimal appliance purchases with respect to energy efficiency. Although the specific policy implications of these respective works are varied, they share the idea that market signals—whether of a free or a government-managed market—shape behavior.

By contrast, the chapter by Archer et al. argues that energy behavior is independent of attitudes and economic considerations. Instead, the authors posit social emulation and learning diffusion as concepts that might have more useful implications for predicting certain kinds of energy behavior. Feldman, in addition, argues that rational models comprehensively misspecify the processes that affect energy-related behavior. Feldman directly assaults the use of the discount rate, a central concept used in economic approaches to studying consumer choices.

On the other hand, market signals cannot be casually dismissed. Price and supply considerations have influenced macro-level energy consequences greatly. Indeed the Hackett study, although small in scale, provides impressive evidence of the impact of market signals in energy consumption—in this case the switch from master to individual metering of energy use. However, Hackett finds that economic variables are only partial explanations of the changes in energy behavior. Fagerson's study of homes in which climate and building characteristics are held constant also suggests that household characteristics will produce substantial variation in energy behavior. Finally, the Vine review of thermostat management studies indicates that no category of variables stands out as a salient influence on thermostat management.

## SOME FINAL ISSUES

During the course of the Summer Study held by the American Council for an Energy Efficient Economy at the University of California, Santa Cruz, a number of issues were raised regarding the interplay between knowledge about energy behavior and public policy. These issues remain apparent in this book. The chapters, all of which are revised from papers originally presented at the Summer Study, suggest complex linkages between information about energy efficiency improvements and behavioral changes. "Facts" about energy efficiency appear to be altered by a host of social factors. Moreover, there is the complexity posed by the tendency for policy options to become merged with other issues, as when local efforts to promote solar energy become ensnared in disputes over the appropriate government role in the economy.

Finally, there remains substantial conflict over the usefulness of rational-economic models of individual behavior. On the one hand, it seems foolish to suggest that individuals do not apply a kind of cost-benefit analysis to such questions as whether to weatherize homes or invest in solar energy. Yet such models often appear highly abstract, so that the task of specifying what, to the individual, comprises a cost or benefit remains, and the need for detailed behavioral data persists. Moreover, the decision-making processes implied by some applications of economic models appears to some as hopelessly inaccurate as a description of how individuals come to select one kind of energy-related behavior over another (e.g., how long to shower or whether to install a solar water heater).

In short, the works included here provide no solace for those who desire a neat blueprint for encouraging energy-efficient behavior. Yet, we feel that they will be of help in ultimately producing policy guides. These works show that we have reached the point at which sound policy planning must incorporate a host of individual variables that mediate between greater energy efficiency and broader social processes.

# SECTION II:

# Consumer Investment Decision Processes

# SECTION II: Consumer Investment Decision Processes

*Willett Kempton*
*Princeton University*

## INTRODUCTION

Why does a consumer purchase an energy-efficient house? Is the motivation one of anticipated energy savings (possibly yielding a rate of return above the mortgage rate or that of alternative investments), a positive attitude toward conservation, or the personal advice of a friend? While admitting that several factors may operate at once, the chapters in this section and those by Baxter and Laquatra in section III analyze and answer the question in different ways.

We have organized the chapters in this section by topic. Two analyze appliance purchases, two examine home retrofit decisions, and two analyze long-term investments.

Both Feldman and Ruderman et al. deal with appliance purchases. Feldman uses appliance data to raise provocative questions about the role of economic factors in consumer energy decisions. He presents data to suggest that consumers are not rational investors. In surveying whether consumers were willing to pay a premium for a more efficient appliance, Feldman found that many respondents fell at the extreme ends of the scale of implicit return on investment. That is, their discount rates were well outside the

*Energy Efficiency: Perspectives on Individual Behavior*

expected range, leading Feldman to conclude that the respondents were not making decisions in the same way an economically rational investor would.

In another survey, he asked the same respondents what they would expect to pay this year to buy what a dollar would have bought a year ago. Theoretically, both answers should have been the same. But the answers were wildly different from each other, and from the actual inflation rate.

Since his data seem inconsistent with economic theory, Feldman proposes that his respondents, and energy consumers generally, can better be understood with reference to psychological principles first outlined by the Swiss psychologist Jean Piaget. Piaget found that children failed to make inferences they were logically capable of, because they were preoccupied with certain salient perceptual features. In particular, Piaget found that children failed to notice that even when an object is physically transformed (such as when clay is molded in the hands), many properties (such as the clay's mass) are "conserved". Feldman speculates that consumers similarly do not necessarily understand that the exchange value of goods is conserved while the nominal value of money is transformed. Rather, they focus on perceptually striking examples, say, of notably large price increases. This focus may lead to their behavior not being economically rational, even if it does make sense from a psychological perspective.

Ruderman, Levine and McMahon, while taking a more standard theoretical approach, perform an innovative analysis of a truly impressive amount of appliance data. They infer the consumer's willingness to purchase higher-priced appliances in return for lower future operating costs. Without taking a stand on how consumers actually make purchase decisions, they attempt to measure the "aggregate market discount rate". This rate characterizes the overall operation of the market, including purchase by third parties, such as plumbers or home builders. In a sense, this rate measures the efficiency of the economy as a whole. While it may not measure the process of any individual's decision, it is useful for Ruderman et al.'s purpose—to evaluate the effect that appliance standards might have on the presently unfettered free market.

Ruderman et al. find high implicit discount rates for energy efficiency, much higher than would be predicted if appliance purchases constituted a rational market. Of the eight appliances studies, only air conditioners—with a discount rate about 20%—showed a rate anywhere near prevailing interest rates. The remainder showed discount rates ranging from 40% to 240%. A unique addition to this analysis is the authors' measure of changes in discount rates through time, using data from 1972, 1978 and 1980. They find that the implicit market discount rates for most of these appliances are not moving closer to interest rates. In fact, many are increasing.

Ruderman et al. propose several causes for this market failure: consumers may not know the relative efficiencies of different models; the appliance may be purchased by someone who will not have to pay fuel costs (e.g., a contractor); prices of energy-efficient models may be inflated by additional ("luxury") features, and finally, there may not be enough high-efficiency appliances available. Whatever the causes, the unusually comprehensive evidence constitutes a strong case for market failure, and that failure represents a challenge both to scientists who might seek to explain it and to policy makers who might seek to correct it.

Wilk and Wilhite, and Archer et al. examine home retrofit decisions. Wilk and Wilhite apply ethnographic methods to the study of California solar and conservation home retrofits. They argue that ethnography, which uses intensive open-ended interviewing to yield detailed contextualized data, can lead to new insights about the processes that generate final consumer choices. As an illustration, they address the anomaly they discovered about attitudes towards weatherstripping. Although weatherstripping would have been a cost-effective retrofit for most of their study houses, few homeowners had installed it, and those who had were not as interested in discussing weatherstripping as their other conservation actions. Most of those who did weatherstrip did so shortly after moving in, as part of "fixing up" the house.

Wilk and Wilhite analyze informant statements, and try to distinguish the rationalizations from the prime causes. First, they found that homeowners think installing weatherstripping is a dirty and undesirable job. Second, they argue that weatherstripping does not fit into normal household categories of maintenance, home improvement, or repair, and it is thus more likely to be avoided or deferred. Finally, energy conservation measures are installed for many reasons apart from energy savings, such as improving a home's aesthetic and resale value, achieving "independence" and increasing comfort. None, save possibly the last, of these additional benefits are realized by weatherstripping. Wilk and Wilhite conclude that even if homeowners are not making the best decisions, analysts must learn how they actually made decisions so that we can improve energy conservation programs.

Archer and his colleagues evaluate alternative models of individual energy decisions. From an earlier analysis of utility energy conservation programs, they infer that the organizers of those programs already accepted one of two models about how individuals make decisions. The first is an attitude model, which is implicit in programs that try to promote positive attitudes about conservation. The other is a rational consumer model, implicit in programs that provide financial incentives for conservation investments. The chapter by Archer et al. uses survey data to test the extent to which these theories, implicit in programs, actually can be scientifically verified as models of consumer behavior.

The authors find surprisingly little evidence for either the attitude or the rational model. Their attitude measures (based on questions such as "How serious do you think this country's energy situation is right now?") showed almost no relationship with self-reported conservation behavior. Perhaps more surprisingly, Archer et al. find little relationship between knowledge of energy programs and conservation actions. For example, households that had purchased solar devices were only slightly more able to name the tax credit percentage (within ± 10%) than households that had not.

Using correlation analysis, Archer et al. show that conservation is not a single behavioral phenomenon. They divide conservation into three categories: practicing conservation habits, purchasing conservation devices, and purchasing solar devices. These three types of behavior have different social determinants. For example, conservation habits are best predicted by self-reported efforts to conserve energy, but not by socioeconomic status. Purchasing both solar and conservation devices is strongly predicted by socioeconomic status, as might be expected by ability to pay. Solar devices are more strongly predicted by the purchaser having personal contacts as sources of energy information.

The authors suggest that energy policies will be more successful if they do not assume either the attitude or rational model of the consumer. They also argue that diverse policies may be more successful than unitary ones in affecting all three types of conservation behavior.

Holt and Rabl deal with situations encountered in long-range conservation investment decisions. Holt analyzes superinsulation, a highly economic technology that is utilized far less frequently than is economically justified. Holt acknowledges explanations previously given for this underutilization, such as lack of knowledge of the benefits and fear of indoor air pollution. He argues that two other factors can explain the low numbers of superinsulated houses. The two factors are high mobility of the US population and lack of confidence by home buyers in recouping their investment. Although data on resale value are sparse, Holt cites figures suggesting low recovery of cost in home resale (see also Baxter et al. in this volume).

Using three cases that vary in level of superinsulation investment, all of which are highly economic in the long run, Holt shows that energy investments benefit a buyer only if she holds the house more than four years, or, in the case of the most complete insulation package, more than twelve years. Since such investment does benefit the entire sequence of buyers—as well as the society as a whole—the resale value problem represents a public goods challenge to policy makers.

Rabl's chapter deals with decisions that must be made under uncertainty about future energy prices and efficiency savings. Many analyses of capital investment calculate the economically optimal level of conservation, given specific assumptions about savings and future energy prices. However, since savings and future prices are highly uncertain, Rabl argues that optimum investment levels should be based on a range of effects over a distribution of possible savings and prices. The question he raises is "How large is the life-cycle cost penalty for too much or too little investment?" He does not claim to describe how consumers actually make decisions. Rabl finds that the optimum conservation curve has a broad minimum. For example, he analyzes one hypothetical case in which the true price of energy is doubled from the price that was guessed when the energy conservation investment was made, yet the total life-cycle cost ratio is increased by only 6%.

Earlier work on energy price uncertainty had suggested that over-conservation was an appropriate strategy. Under one of the two models Rabl analyzes, over-conservation is not always the best strategy. Since this model may be less general, and since investors may want to reduce risks associated with future rises in energy prices, Rabl concludes that conservation overinvestment in the face of uncertainty may be a good choice in practice.

# Why Is it So Hard to Sell "Savings" As A Reason For Energy Conservation?

*Shel Feldman*
*National Analysts*[1]

## INTRODUCTION

In the 1982 ACEEE Summer Study, Awad and I (Feldman and Awad 1982) identified four attitudinal factors that relate to intentions for conserving electricity. These four factors are:

- lack of cynicism—the belief that conservation efforts make a difference and that individuals can control their wants, their energy use, and their energy costs;
- concern for supply needs;
- concern for what is socially acceptable regarding electricity conservation (fulfilling social norms); and
- concern for monetary savings.

In passing, I note three things: First, we did not find any general conservation ethic among members of our sample. Second, perceptual and attitudinal variables are linked to intentions to conserve electricity; the objective factors we studied are not sufficient to explain those intentions. Third, concern with monetary savings not only failed to provide a full explanation of the variance in intentions, but it was also not a significant predictor in the case of

*Energy Efficiency: Perspectives on Individual Behavior*


[1] A division of Booz-Allen and Hamilton, Inc.

most specific intentions considered, and it was not the major explanatory variable of our analyses.

**Table 1. Regression Weights for Predicting Intentions to Conserve Electricity, by End Use Considered.**

| | Predictor | | | | | |
|---|---|---|---|---|---|---|
| End Use | Beliefs | Cynicism | Concern for Supply | Avowal of Social Norms | Home Econ. | $R^2$ |
| Air Conditioner | 0.318* | -0.104* | -0.078 | 0.060 | 0.069 | 0.11 |
| Washer/ Dryer | 0.223* | -0.055 | 0.056 | 0.113* | -0.015 | 0.07 |
| Dishwasher | 0.387* | -0.106* | -0.003 | 0.062 | 0.041 | 0.17 |
| Heating | 0.291* | -0.026 | 0.058 | 0.095* | 0.069* | 0.12 |
| Lights | 0.148* | -0.069* | 0.072* | 0.200* | 0.055 | 0.11 |

* statistically significant ($p < 0.05$ at the 95% confidence level)

We and others connected with a project sponsored by Southern California Edison have reported additional data on attitudinal concommitants of conservation intentions (Awad et al. 1983) and expanded on the model involved (Feldman et al. 1983; Williams 1983).[2]

Expansion of the model was needed for two reasons: First, no psychological model I know of claims that behavior is a direct function of attitudes, except in the most extremely constrained laboratory situation. As Williams (1983) has suggested, following Fishbein and Ajzen (1975), voluntary actions depend on both behavioral intentions and opportunities to perform the behaviors at issue. In turn, behavioral intentions are affected both by preferences and by the perceived normativeness of the behaviors at issue; and preferences are affected both by beliefs about the attributes of

[2] The author thanks Southern California Edison for permission to cite these data. The opinions expressed in this chapter are those of the author, however, and do not necessarily represent the opinions or policies of the Southern California Edison Company.

the behavior and attitudes relating to those attributes. Hence, the tion, housing, and associated life-style preferences.

Second, for a variety of practical reasons, it is inappropriate to depend heavily on a strictly attitudinal model in demand-side planning. Education toward energy conservation is only one of several demand-side strategies. Among others are hardware-based strategies, such as obtaining consumer agreement to the application of direct load control and promotion of energy-efficient housing and appliance stock; other behavioral strategies, such as fostering changes in time of use in order to modify load curves; and economic strategies, such as overall price increases, price increases for selected fuels or selected times or amounts of use, and changes in rate structures.

Indeed, strategies focusing on behavioral change through persuasion and education suffer from particular difficulties, including the fact that conflicting motives constantly recur, since many changes are inconvenient and cause discomfort to the individual, and many require cooperation among various members of the family unit. Moreover, current energy supplies appear quite adequate, with price rises having moderated significantly and little public attention being devoted to the issue. Last—but not least—neither public agencies nor utility companies have a great deal of control over relevant sources of communication and persuasion: Compare the total public information budget of any utility or group of utilities with the amount spent on promoting a new additive to an existing soap in the hope of gaining two or three share points.

One approach we and others have attempted, given the factors discussed above, has been to expand the range of studies of the determinants of energy conservation attitudes, intentions and relevant behaviors. In particular, we have recently attempted to measure the amount of investment in energy efficiency reported by ratepayers, both in particular technologies and overall, the amount of investment intended, and the willingness to purchase energy-efficient technologies at some premium over less efficient technologies.

This last variable—willingness to pay some premium for energy-efficient technology—has seemed rather promising, both for practical and for theoretical reasons. Practically, if consumers can be induced to purchase energy-efficient housing and appliances, the difficulties cited earlier with regard to persuading people to engage *repeatedly* in energy-conserving behavior could be avoided. Energy savings could be built into our society, and they could presumably be incorporated without government mandates and coercion. Theoretically, we would be able to hook into the powerful choice models developed by economists such as Hausman (1979), treating attitudes and preferences as simply additional variables to be monetized and included in the analysis of the implicit discount rate. Moreover, by working with these variables, we are clearly making use of and expanding on the concern for monetary savings that was earlier identified as one of the psychological factors that motivates people to conserve.

## SAVINGS AND THE IMPLICIT DISCOUNT RATE

The implicit discount rate approach is based on the postulate that the rational person offered several investment options prefers that option with the lowest life-cycle costs. Since life-cycle costs include first costs, operating and maintenance costs, and interest costs, it is simple to compare the costs of two appliances, say, whose performance characteristics are equal except for energy efficiency.

In many cases, the first costs for the more energy-efficient unit are higher than those for the less efficient unit, but the operating and maintenance costs of the more energy-efficient unit are lower. Hence, a larger initial investment would result in later savings, which should equal, and then exceed, the initial price difference at some time in the future. By determining how soon the energy-efficient unit must "pay back" the differential initial investment, the analyst can determine the interest rate that the buyer implicitly requires on his or her initial investment. The shorter the payback period required, the larger the implicit discount rate, and the less willing the individual is to invest in energy efficiency.

With this model, we can also ascertain the effects of other differences in performance characteristics on implicit discount rates, as well as the effects of different demographic characteristics of ratepayers, and the effects of different attitudes and beliefs. That is, we can determine whether different groups of ratepayers, such as younger and older people or homeowners and renters, differ in their implicit discount rates, and whether such motivational factors as cynicism or concern for monetary savings are associated with different discount rates.

The results of these analyses could be used in various ways. For example, they could be used to target subsidy programs—attempting to provide incentives that are most cost-effective to induce purchase of energy-efficient technology, to just the right peo-

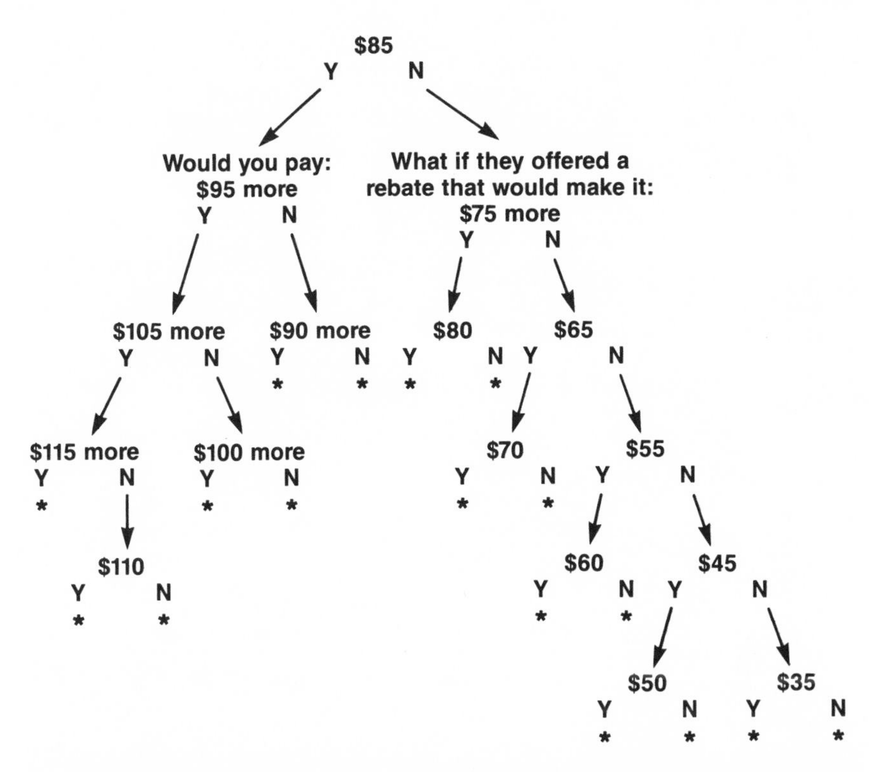


**Fig. 1. Item wording for measurement of implicit discount rate. Suppose you needed a new refrigerator *now*, and you decided to buy a 19 cubic foot model. As you may know, more energy-efficient models cost a bit more than less energy-efficient models, but all refrigerators tend to last at least 10 years without repairs. If your electric company offered you a model that would save $25 a year in energy costs, would you pay $85 more to purchase it in the first place? (*marks the end of question.)**

ple, in just the right amount.

This model poses some difficulties, however. It assumes a rational investor, with perfect knowledge of the marketplace, costs, benefits, and preferences. Of course, economists and others recognize that perfect knowledge seldom exists, and a number of proposals have been made to deal with the problem. Among these are rules for labeling houses and appliances with energy cost information and enhancing the visibility and clarity of the labels.

There is at least some reason to doubt that most consumers do in fact use the information presented in any comprehensive way. If people act as if they are computing required payback periods or discount rates on the basis of perceived costs, benefits and prefer-

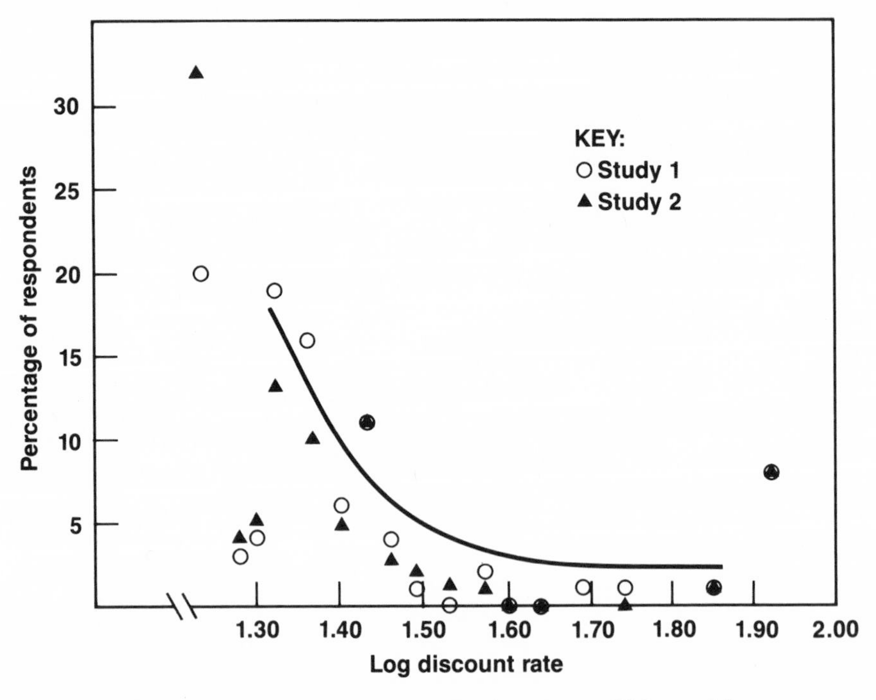


**Fig. 2. Distributions of implicit discount rates for energy-efficient refrigerators.**

ences, we might expect that observed discount rates are more or less normally distributed across people. After all, perceptions of costs, benefits and preferences differ more or less normally across people, and their effects should be expected to combine in such a way as to create a randomly distributed implicit discount rate. Various constraints in the system could be expected to skew the observed distribution, or to cause it to become more leptokurtic—that is, taller and thinner—than normal. For example, if people observed available money market interest rates for individual investment and set those as the minimum acceptable discount rates for energy-efficiency investments, the distribution might be positively skewed, and rather leptokurtic just above the market rate, but still a recognizable distortion of a normal distribution.

In two studies conducted for Southern California Edison (Feldman et al. 1983), we asked respondents to a telephone survey to indicate the amount of additional money they would pay for an energy-efficient refrigerator that would last at least 10 years, and would save $25 per year in energy costs. (See Figure 1 for the exact wording and flow of the questions.) The distributions of implicit discount rates are shown in Figure 2, and the actual distributions of responses are presented in Table 2.

We have taken the liberty of drawing in a smooth curve in Figure 2 that more or less represents the points to an implicit discount rate of about 21 percent or more (assuming a 10-year refrigerator life). And we can argue that there are some end effects that plague our analysis, just as they affect that of Houston (1983). Something is clearly wrong, however: The distribution is in no way smooth, and it is not simply a matter of one study, or the numbers of respondents included. We find our analyses improve significantly when we abandon our ratio scale and simply categorize respon-

**Table 2. Distribution of Discount Rates for an Energy-Efficient Refrigerator**

| Maximum Acceptable Addition To Purchase Price | 10-Year Discount Rate[1] | Study 1 (N=100) Percentage | Study 1 (N=100) Percentage At Given Rate Or Lower | Study 2 (N=1248) Percentage | Study 2 (N=1248) Percentage At Given Rate Or Lower |
|---|---|---|---|---|---|
| (Total) | | (100) | (97) | (100) | (98) |
| $30 or less | 83 | 8 | 97 | 8 | 98 |
| $35 | 71 | 1 | 89 | 1 | 90 |
| $45 | 55 | 1 | 88 | * | 89 |
| $50 | 49 | 1 | 87 | 1 | 89 |
| $55 | 44 | 0 | 86 | * | 88 |
| $60 | 40 | 0 | 86 | * | 87 |
| $65 | 37 | 2 | 86 | 1 | 87 |
| $70 | 34 | 0 | 84 | 1 | 86 |
| $75 | 31 | 1 | 84 | 2 | 85 |
| $80 | 29 | 4 | 83 | 3 | 83 |
| $85 | 27 | 11 | 79 | 11 | 80 |
| $90 | 25 | 6 | 68 | 5 | 69 |
| $95 | 23 | 16 | 62 | 10 | 64 |
| $100 | 21 | 19 | 46 | 13 | 54 |
| $105 | 20 | 4 | 27 | 5 | 41 |
| $110 | 19 | 3 | 23 | 4 | 36 |
| $115 or more | 17 | 20 | 20 | 32 | 32 |
| Missing | -- | 3 | -- | 2 | -- |

* Less than 0.5 percent

[1] Assuming $25 return per year with no salvage value and immediate replacement.

dents into those with low, medium and high discount rates such as those utilized by Hausman (1979). People appear to respond far more grossly than the implicit discount rate model seems to imply.

Thus, while we have some reservations about the methods used by Houston (1983), we concur with one form of his basic conclusion: A substantial number of people do not appear to behave in accordance with the model underlying the implicit discount rate approach. In his analysis, Houston found those with lower incomes and less experience with energy-conserving activities less likely to respond to his discount rate question. Hausman found that those with lower incomes acted as if they had extremely large discount rates, and Gately (1980) reported truly enormous discount rates for many in his sample—again calling into question whether the model is appropriate as a description of the behavior of the individuals studied. In our study, we found that implicit discount rates varied with the future orientation of the respondents, as well as with their reported experience with energy-efficiency investments. Furthermore, 28 percent to 40 percent of the responses were off the ends of the scale.

Why don't people behave in accordance with the model? Houston suggests that many do not have the appropriate conceptual tools, but he does not describe what those tools are, and why some may lack them. We suggest that there is a basic lack of ability to impute value to an object and to conserve that value in the face of apparent changes over time. In the remainder of this paper, we shall endeavor to explain the *psychological* notion of conservation, indicate its relevance to economic concepts such as value, and suggest some implications of this analysis for problems in energy conservation.

## CONSERVATION IN PSYCHOLOGICAL THEORY

Jean Piaget, a well-known Swiss psychologist, first described an experiment with young children (Piaget & Inhelder 1941) that has since been replicated thousands of times. The adult takes two lumps of clay and rolls each into a ball, adding to or removing clay from one, until the child agrees that each contains the same amount of clay. The adult then rolls one into a sausage form, as the child watches, and asks whether the two pieces *now* contain the same amount of clay, or one has more, or one has less than the other. Typically, the child of five or six will say the two pieces are no longer equal. (Most often, the child seems to focus on the length of the "sausage" and says it contains more clay than the ball.) Piaget explains the phenomenon as indicating that the child focuses on one salient dimension in the transformation from ball to sausage and, seduced or overwhelmed by the difference between the two pieces of clay on that one dimension, fails to recognize the concomitant shift in other dimensions such that the amount of clay has been conserved during the transformation.

Of particular interest, the child is usually perfectly capable of recognizing that no clay was added or taken away during the transformation, perfectly capable of noting that the two pieces of clay are equivalent when the sausage is returned to the form of a ball. Indeed, it is only when the adult asks for a *logical* explanation that the child engages in any distortion of reality—asserting that the adult must have *secretly* added clay to the "sausage," for example. In other words, the child is not necessarily lacking in the logical knowledge that the clay cannot have become more or less in being reshaped; the child simply cannot assimilate that knowledge to the perceptual situation with which he or she is confronted. However, the child will continue to insist that there is more clay in one piece than another when the shapes differ; and no amount of logical argument or adult remonstrance, and no direct tuition, changes the child's mind.

Only with additional maturity and with what Piaget calls experience with accomodation and assimilation of various schemata does the child come to "conserve" changes in mass while objects undergo shape transformations.

This demonstration is but one of a class of similar phenomena. Piaget and his followers have shown that children have difficulty conserving various relationships in the face of different transformations. Among these are the physical relationships including mass, number, and volume, but also such non-physical relationships as family membership.

Moreover, while the preschooler may have trouble with conservation in the face of direct physical transformations, older children encounter similar difficulties in dealing with symbolic transformations of greater complexity. A classic example of a problem posed to adolescents is the following: "Edith is fairer than Suzanne. Edith is darker than Lilly. Who is darkest of the three—Suzanne, Edith, or Lilly?" Solution of this problem requires constructing the inverse of one of the statements given, while conserving the other existing relationships in the face of the symbolic transformations.[3]

According to Piaget, the structure of the problem facing the subject of the symbolic transformation experiment is the same as that facing the child watching the manipulation of the balls of clay; where perceiving the length of the "sausage" may dominate the ability to appreciate the reciprocity of the change in diameter to the change in length, the explicit verbalizations of "darker" and "lighter" now dominate the reciprocity of the logical relationships. Only the specific relationship and the level of symbolic abstraction have changed.

What has all this to do with value, investments, and implicit discount rates? Simply this: The model underlying the computation and study of implicit discount rates assumes that people readily compute the present value of an investment and readily project that value into the future. We suggest that such computations and projections are extraordinarily complex, and require a conservation of value that is unlikely to be in ordinary use by many people.

In developing investment strategies keyed to internal rates of return, we assume that an individual utilizes two types of knowledge. The first is straightforward: the current market rate of interest, modified by expectations about changes in that rate over the effective life of the investment contemplated. But the second may not be as readily applied as we are prone to assume.

Specifically, the individual must first *recognize* the need to project into the future the relationship between the cost of the goods or services in which investment is contemplated and his or her current preferences. In doing so, the individual must also recognize the dependence of that relationship on the value of money and on his or her preferences. But these are complex relationships, and not everyone appreciates them, or even the need to understand them.

## MONEY AND ITS FUNCTIONS

Economists traditionally treat money as serving two major functions. First, it serves as a medium of exchange that facilitates the transfer of goods and services among producers and consumers.

[3] Thus, the child might construct the inverse of the first statement, obtaining "Suzanne is darker than Edith." Only now can she solve the problem using transitivity ("Suzanne is darker than Lilly").

To the extent that money is held over time and exchange is deferred, it also serves to store value. But a full understanding of the use of money in this function requires appreciation of the opportunity costs of saving, the origin of interest payments, and the relationships between liquidity preferences and investment behavior.

In its second major function, money serves as a standard of value, without which a complex society cannot function. It is only as price ratios are translated into common units that the value of the diverse products of an industrial society can be measured against one another, and only as money is integrated with labor, time, effort and skill, in the wage-unit, that various forms of work can be compared with one another. Moreover—and this issue is the critical one for our present purposes—it is only to the extent that the delivery of future products or future labor can be valued, with money serving as a standard of deferred payment, that contracts can be written and enforced, and that future value can be compared with present value in terms of payback periods, implicit discount rates, or avoided costs.

But there is ample evidence to suggest that many people do not understand how money functions as a standard of deferred payment. In a recent survey for the Public Opinion Index, for example, we sought to study public perceptions of inflation. In telephone interviews, we asked 1004 persons in a national probability sample the rate of inflation in 1983. The median estimate among the 44 percent of our sample that answered the question was that inflation ran at 7.75 percent. The median estimate varied among subgroups: for example, from 7.30 percent among professionals, managers, and owners, through 8.24 percent among white-collar and sales-clerical workers, to 8.56 percent among blue-collar workers. But in none of the demographic analyses was the median below six percent. The actual inflation rate in 1983 was slightly over three percent, as measured by the consumer price index.

We also asked members of our sample how much they would have to spend today to get what one dollar bought a year ago, in the belief that they would be more accurate on this more concrete sort of question. Table 3 shows that the exact opposite occurred. While respondents felt capable of answering the question (only 19 percent failed to try to do so), the answers were much further from the true rate of inflation in 1983 than were the direct estimates of inflation rates. The overall median estimate was an astounding $1.41. Differences among subgroups were often quite striking, as

**Table 3. Median Estimates of Money Needed Today to Buy What One Dollar Bought a Year Ago**

| Selected Segment | Median |
|---|---|
| **Total Public** | $1.41 |
| **Sex** | |
| Male | $1.27 |
| Female | $1.60 |
| **Age** | |
| 19 – 24 years of age | $1.41 |
| 25 – 34 years of age | $1.36 |
| 35 – 44 years of age | $1.39 |
| 45 – 64 years of age | $1.45 |
| 65 years of age or older | $1.42 |
| **Educational Attainment** | |
| High school incomplete or less | $1.53 |
| High school graduate | $1.47 |
| College incomplete | $1.29 |
| College graduate or more | $1.20 |

well. The medians vary by $0.33 with level of educational attainment, for example. Even among college graduates, the median was $1.20.

We suggest that people are somewhat closer to reality when asked the inflation rate directly because they hear or read news reports of official figures relatively often. Clearly they do not recall these figures with great accuracy, but at least their guesses are within an order of magnitude. However, people clearly do not integrate the figures they hear or those they recall with their perceptions of how inflation affects them directly–the cost of goods today relative to a year ago. And people clearly do not hold constant in their minds those relative costs. Most people are not using money as a *standard* of deferred payment; they are not conserving the exchange value of goods they are purchasing in the face of changes in the nominal value of money. Rather, like the child with the "sausage," they are focusing on the perceptually arresting transformations that are occurring.

## CONCLUSION

If most people do so poorly in conserving value over time, it is unlikely that they will do well in *projecting* value into the future. Implicit discount rates appear to be distributed categorically in the general public because they reflect the different attitudes of different population segments toward investing in energy efficiency. Avoided costs and implicit discount rates are probably not useful concepts for describing the behavior of the general public, however useful they may be for analyzing the behavior of commercial and industrial decision makers. We would do well to avoid building our models and preparing our advertising campaigns based upon the assumption that the energy consumer operates—or can operate—as a rational investor.

**References**

Awad, Z.A., R.H. Johnston, Jr., S. Feldman, and M.V. Williams. "Customer Attitude and Intentions to Conserve Electricity." *Advances in Consumer Research* 10:652-654, 1983.

Feldman, S. and Z.A. Awad. "What Appeals Should We Use to Promote Electricity Conservation?" Presented at the ACEEE Summer Study on Energy Efficiency in Buildings, Santa Cruz, California, August 1982.

Feldman, S., Z.A. Awad, and M.V. Williams "Cost Consciousness, Credit Use and Future Orientation Factors in Reported Family Conservation Investment Behavior and Discount Rates." In *Abstracts and Papers, Families and Energy: Coping With Uncertainty*, Michigan State University, October 1983.

Fishbein M., and I. Ajzen. *Belief, Attitude, Intention and Behavior: An Introduction to Theory and Research.* Reading, MA: Addison-Wesley, 1975.

Gately, D. "Individual Discount Rates and the Purchase and Utilization of Energy-Using Durables: Comment." *Bell Journal of Economics.* 1980, 11, pp. 373-374.

Hausman, J.A. "Individual Discount Rates and the Purchase and Utilization of Energy-Using Durables." *Bell Journal of Economics.* 10: pp. 33-54, 1979.

Houston, D.A. "Implicit Discount Rates and the Purchase of Untried, Energy-Saving Durable Goods." *Journal of Consumer Research.* 10:236-246, 1983.

Piaget, J., & B. Inhelder. *Le developpement des quantites chez l'enfant.* Neuchatel: Delachaux et Niestle, 1941.

Williams, M.V. "Segmenting Markets for Conservation." *Proceedings of Utility Conservation Programs: Planning, Analysis and Implementation*, 1983.

# Energy-Efficiency Choice in the Purchase of Residential Appliances

*Henry Ruderman, Mark Levine, and James McMahon*
*Lawrence Berkeley Laboratory*

## INTRODUCTION

This chapter provides a quantitative analysis of the behavior of the market for the purchase of energy efficiency in residential appliances and heating and cooling equipment. Accurate forecasts of residential energy use require quantitative assessments of market decisions about energy efficiency. The results of our investigation of market behavior can lead to a better understanding of the barriers to investment in energy conservation. Understanding market behavior over time is a prerequisite to an evaluation of the need for and the importance of policies to promote energy efficiency.

In this chapter, we examine the historical efficiency choices for eight consumer products: gas central space heaters, oil central space heaters, room air conditioners, central air conditioners, electric water heaters, gas water heaters, refrigerators, and freezers. These products were selected because they account for a major part of residential energy consumption, data on efficiency and costs are readily available, and they are under consideration by DOE for efficiency standards.

We characterize the behavior of the market for these eight products by a single quantity that we call an aggregate market discount rate. The aggregate market discount rate quantifies the behavior of the market as a whole with respect to energy efficiency decisions. Choices by individual purchasers are constrained by the decisions made by the manufacturers of appliances, the wholesalers or retailers who distribute them, and the third-party appliance installers such as builders or plumbers. The value of the discount rate reflects the actions of all these decision makers. It is determined empirically from data on the efficiency and cost of appliances purchased between 1972 and 1980. By examining the historical behavior of the market discount rate, we can better understand the factors that influence efficiency choice. Furthermore, the market discount rate can be used as a parameter in forecasting future residential energy consumption. More detail about this work may be found in a Lawrence Berkeley Laboratory report (Ruderman et al. 1984).

## METHOD

A discount rate is a measure of the present value of money received or spent in the future. For example, if someone values an income of $110 received a year from today the same as an income of $100 received today, that person has a discount rate of 10 percent per year. Thus, given the discount rate $r$, one can calculate the present value of a stream of income (or expenditures) using the formula

$$PV = \sum_{t=1}^{N} \frac{X_t}{(1 + r)^t},$$

where

$X_t$ = Income in year t

and

$N$ = Duration of income stream in years.

For a constant stream of income, this formula becomes

$$PV = PWF \cdot X_t,$$

where we have defined the present worth factor by

$$PWF = \sum_{t=1}^{N} \frac{1}{(1 + r)^t} = \frac{1}{r} \left[ 1 - \frac{1}{(1+r)^N} \right]. \tag{1}$$

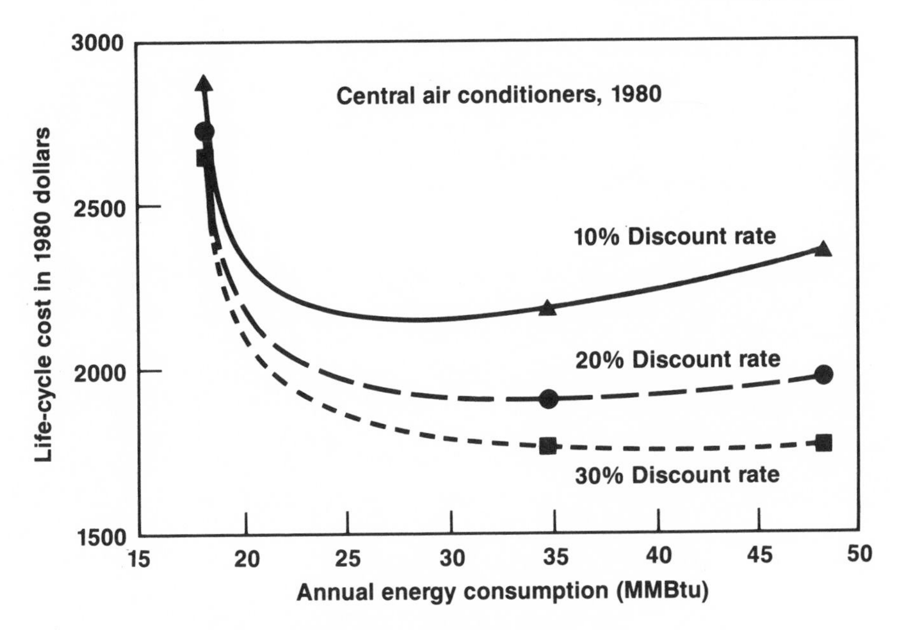


**Fig. 1. Life-cycle cost curves as a function of discount rate.**

The analysis assumes that the behavior in the appliance marketplace can be characterized as if the purchasers of appliances minimize the life-cycle cost of owning and operating them. The observed average efficiency choice is characterized by an aggregate market discount rate, which is the discount rate calculated when the observed average efficiency choice is assumed to lie at the minimum of the life-cycle cost curve. Even if purchasers do not actually decide on the basis of life-cycle costs, the market discount rate is useful as a measure of market imperfections. This analysis examines consumer choice for products with different efficiencies that use the same fuel, and does not account for other factors that might influence consumer choice.

Figure 1 illustrates the effect of different discount rates on the position of the minimum of the life-cycle cost curve. At higher discount rates, the slope of the operating cost component is lower, and the minimum is at higher annual energy consumption and lower appliance efficiency. For central air conditioners, a discount rate of 20 percent puts the minimum of the LCC curve at 34 million Btu, corresponding to the annual energy consumption during 1980. This point is marked on all three curves. Rather than using the discount rate to locate the minimum of the life-cycle cost curve, we reverse the process and determine the market discount rate from the position of the minimum, which is assumed to occur at the average energy use.

The life-cycle cost for an appliance is the sum of the purchase cost and the discounted operating cost:

$$LCC = PC + PWF \cdot FP \cdot TI \cdot E, \tag{2}$$

where PWF is the present worth factor defined above, FP is the average fuel price (assumed constant over time, i.e., the consumer expects no price escalation), TI is the relative thermal integrity, and E is the average annual energy consumption by the appliance. The thermal integrity factor is included for temperature-sensitive appliances to account for the effects on energy consumption of changes in the thermal characteristics of the building shell (such as improved insulation). Finding the minimum of Equation 2 with respect to energy and solving for the present worth factor gives determine the aggregate market discount rate.

For the purchase cost vs. energy-use relationship we fit the data to an exponential curve of the form

$$E = E_\infty + (E_o - E_\infty) \exp[-A(C - 1)]. \tag{4}$$

where

| | | |
|---|---|---|
| $E$ | = | annual unit energy consumption (UEC) |
| $E_o$ | = | highest UEC on cost vs. energy-use curve |
| $E_\infty$ | = | minimum UEC attainable at infinite purchase cost |
| $C$ | = | PC/$PC_o$ |
| $PC$ | = | purchase cost corresponding to E |
| $PC_o$ | = | purchase cost corresponding to $E_o$ |

and A is a parameter determined from the shape of the curve. Using this expression, the present worth factor becomes

$$PWF = \frac{PC_o}{A \cdot FP \cdot TI} \frac{1}{(E - E_\infty)}. \tag{5}$$

The data required to perform an analysis of aggregate market behavior include: 1) purchase price and unit energy consumption of alternative design options for each product; 2) average efficiency of the appliances purchased; 3) energy prices; 4) thermal characteristics of houses; and 5) average appliance lifetimes.

The major sources of data on the costs and energy use of appliances are the engineering and economic analyses performed for the

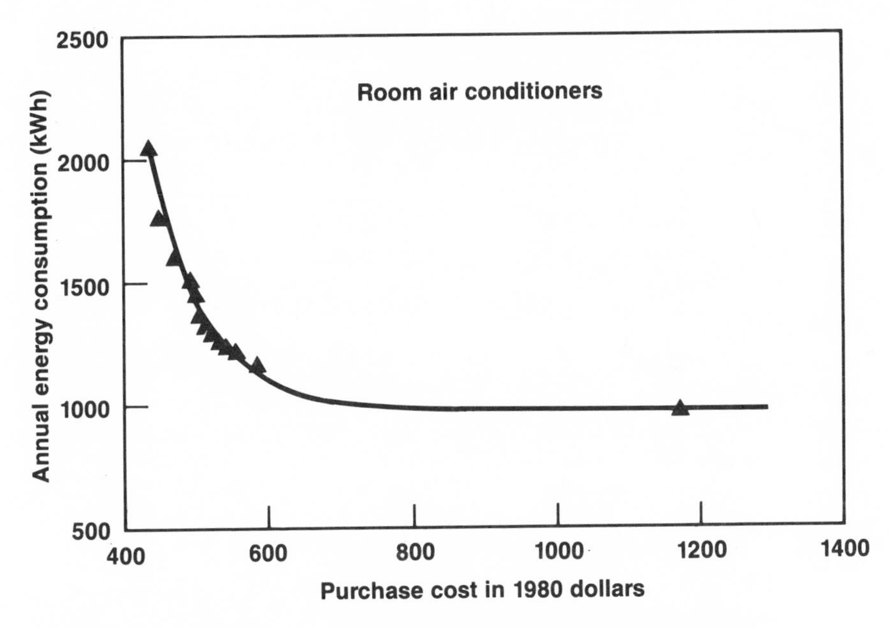


**Fig. 2. Cost and energy use of design options for room air conditioners.**

U.S. Department of Energy analysis of Consumer Product Efficiency Standards (U.S. DOE 1982 a and b). These reports provided estimates of the purchase prices of individual appliances with different efficiencies, as well as estimates of their average usage. Supplemental data were obtained from Arthur D. Little, Inc. to extend the data back in time to 1972, and forward to 1980 from the original data sets for 1978. The data were aggregated from the various appliance classes into a single set of data points representing the product type. Finally, a least-squares fit was performed to the functional form specified above to obtain the parameters of the curve. A typical cost vs. energy-use curve is shown in Figure 2.

The U.S. DOE CS-179 Survey of Manufacturers (U.S. DOE 1982b) provides historical data from appliance manufacturers on the average efficiencies of units shipped in 1972 and 1978. The efficiency factor (e.g., efficiency, energy efficiency ratio, etc.) for each model is multiplied by the number of units of the model shipped in that year, then summed over all models and divided by the total shipments to give the shipment weighted energy factor (SWEF). Table 1 shows the SWEFs used to calculate the market discount rates. Efficiency data from trade associations and individual manufacturers were used to check our results.

**Table 1. Shipment weighted energy factors (SWEF).**

| Appliance | 1972 | 1978 | 1980 |
|---|---|---|---|
| Gas Central Space Heater (AFUE %) | 62.7 | 63.6 | 65.9 |
| Oil Central Space Heater (AFUE %) | 73.6 | 75.0 | 76.0 |
| Room Air Conditioner (EER) | 6.22 | 6.75 | 7.03 |
| Central Air Conditioner (SEER) | 6.66 | 6.99 | 7.76 |
| Electric Water Heater (%) | 79.8 | 80.7 | 81.3 |
| Gas Water Heater (%) | 47.4 | 48.2 | 51.2 |
| Refrigerator ($ft^3$/kWh/day) | 4.22 | 5.09 | 5.72 |
| Freezer ($ft^3$/kWh/day) | 8.08 | 10.07 | 10.83 |

Source: Department of Energy Survey of Manufacturers (U.S. DOE 1982b)

**Table 2. Aggregate market discount rates for appliances (in percent).**

| Appliance | 1972 | 1978 | 1980 |
|---|---|---|---|
| Gas Central Space Heater | 33.5 | 41.9 | 45.1 |
| Oil Central Space Heater | 42.8 | 58.9 | 85.1 |
| Room Air Conditioner | 17.9 | 19.5 | 17.3 |
| Central Air Conditioner | 17.1 | 21.8 | 16.1 |
| Electric Water Heater | 209.1 | 244.4 | 243.2 |
| Gas Water Heater | 66.5 | 93.4 | 102.0 |
| Refrigerator | 74.0 | 69.0 | 59.2 |
| Freezer | 167.4 | 148.8 | 138.2 |

Average energy prices for the three years were obtained from the Energy Information Administration (U.S. DOE 1979). Winter and summer marginal electricity rates are used for heating and cooling equipment, respectively. Marginal rates are calculated as the average rate for the 500 to 1000 kWh per month block. Thermal integrity factors were defined as the relative annual energy consumption for space conditioning end uses, reflecting changes from the thermal characteristics of the stock house in existence in 1977 (including changes in insulation, window glazings, infiltration, etc.). Historical integrity values were estimated from survey data (U.S. DOE 1980). The appliance lifetimes are the same values used in the Consumer Product Efficiency Standards analysis (U.S. DOE 1982a).

## RESULTS

Aggregate market discount rates were calculated for years 1972, 1978 and 1980. The first step is to estimate the parameters of the annual energy use vs. purchase cost curve for each appliance. Evaluating Equation 5 at E corresponding to the SWEF gives the present worth factor, which is then converted to a discount rate. The results presented in Table 2 are based on a single cost-efficiency curve for each appliance covering the period 1972-80. The tabulated discount rates are expressed in percent per year. Changes in discount rates over time for a single appliance are due to changes in SWEF, fuel prices, and, in the case of temperature-sensitive appliances, thermal integrity. The observed discount rates range from less than 20 to more than 200 percent per year. Those for central space heating and water heating appear to be increasing over the time period, whereas the others are either decreasing or remaining constant.

To understand these year-to-year differences, we performed an analysis of the sensitivity of the results to changes in SWEF and the cost-curve parameters. The observed discount rate is extremely sensitive to the assumed SWEF; a change in SWEF of less than five percent could explain the year-to-year differences. Of the parameters of the cost-energy use curve, the greatest sensitivity is to the value of $E_\infty$. Since the other three parameters enter Equation 4 in a similar way, they all have the same percentage effect.

It is our judgement that the SWEFs are known to within five percent. They would thus lead to relatively little uncertainty in the discount rate. The parameters of the cost-efficiency curve are less well known, perhaps to within ten percent. Because of the fitting procedure, $E_\infty$ may be too low, leading to a high discount rate, possibly by as much as 20 percent. Future work on the cost and efficiency of the most efficient products should lead to better estimates of these parameters. We do not believe that the uncertainty will affect the observed changes in discount rates over time.

In calculating the discount rates, we assume that purchasers do not anticipate any escalation in real fuel prices. This assumption may not be warranted, but it is a conservative one. Putting an assumed inflation rate for energy into the calculation would result in higher values for the observed discount rates.

As a check on our results, market discount rates were calculated for refrigerators, freezers, gas furnaces, and room and central air conditioners using historical data on efficiencies from several additional sources. The discount rates show the same trends as the CS-179 data.

The high discount rates observed in this study are difficult to interpret as the result of the operation of a rational market. If a consumer's discount rate is higher than the current interest rate and if prices reflect costs, a rational consumer would borrow money to purchase a more efficient appliance. For example, an investment of $21 to include increased door insulation, a higher compressor efficiency, a double door gasket, and an anti-sweat heater switch in a refrigerator would save $22 per year at 1980 fuel prices, an annual return of 105 percent on the investment. The data, however, indicate that these investments are not prevalent. We believe that the high discount rates show that imperfections in the market prevent consumers from making economically optimal decisions because:

- consumers may not have adequate information about appliance efficiencies or access to capital markets;
- the person purchasing the appliance may not be the one who uses it;
- the price of the appliance may be determined by factors other than efficiency;
- or high-efficiency appliances may not be produced in large enough quantities to satisfy the demand.

Thus all the participants in the marketplace could contribute to making the discount rate high.

## CONCLUSIONS

Several generalizations may be made from the basic results: (1) the values of the aggregate market discount rate for the appliances studied are higher than real interest rates or the discount rates commonly used in life-cycle cost analyses of consumer choice; (2) the aggregate market discount rates appear to be relatively constant over time, with rates for some products (space and water heating) increasing somewhat over the past decade and rates for others (freezers and refrigerators) decreasing over the same time period; and (3) the sensitivity analyses show considerable changes in results as inputs are varied. This large variation combined with other limitations of the analysis suggests that considerable care must be used in discussing the numerical results; however, the first two observations are likely to be meaningful in a qualitative sense.

Overall, the high values of the aggregate market discount rates in Table 2 indicate that the average appliance or heating and cooling system purchased does not include energy efficiency measures that yield very high returns on investment. Several different explanations of the phenomena of underinvestment in energy efficiency in the residential sector have been proposed:

- lack or high cost of information about costs and benefits of energy efficiency improvements or a lack of understanding by purchasers of how to use this information if it is available;
- the prevalence of indirect or forced purchase decisions (e.g., landlord purchase of equipment for rental property; need for immediate replacement of malfunctioning equipment);
- unavailability of high efficiency equipment in retail stores or the unavailability of highly efficient equipment without other features (so-called "gold-plating") that may not be desired by the average purchasers;
- manufacturer's decisions to improve product efficiency are often secondary to other design changes and take several years to implement; and/or
- marketing strategies by manufacturer or retailer that intentionally lead to sales of less efficient equipment.

Several studies have been initiated to compare these explanations with empirical data.

A significant finding from Table 2 is that the aggregate discount rates have changed only modestly over time. We are aware of no other work that has investigated the behavior of the market for energy efficiency in residential appliances over time. Previous studies estimate discount rates for a single appliance type during one year (Hausman 1979; Gately 1980; Meier and Whittier 1982). Our work indicates that the behavior of the market during the period 1972 to 1980 has been relatively unchanged (in terms of return on investment for energy efficiency in consumer products). The market for appliances does not appear to be influenced by rapidly rising energy prices and consumer awareness of energy issues. This conclusion is similar to the results for investment in thermal integrity in houses obtained by Levine and Scott (1986).

Several investigations into the behavior of the appliance marketplace are underway to follow up on this work. We are continuing to examine the possible effects of several programs aimed at making efficient appliances more attractive to purchasers, including research on utility incentive programs and the FTC labeling program. Future analysis will consider alternative methods for characterizing market behavior, such as internal rates of return and payback periods.

**References**

Gately, D. "Individual Discount Rates and the Purchase and Utilization of Energy-Using Durables: Comment." *Bell Journal of Economics* 11(3):373, 1980.

Hausman, J.A. "Individual Discount Rates and the Purchase and Utilization of Energy-Using Durables." *Bell Journal of Economics* 10(1):33, 1979.

Levine, M.D. and R.E. Scott. "Estimates for an Economic Model Incorporating Price and Usage Elasticity Adjustments for Consumer Decision-Making over Energy Efficiency Options in the Purchase of New Single-Family Housing." Lawrence Berkeley Laboratory report, in preparation.

Meier, A. and J. Whittier. "Purchasing Patterns of Energy-Efficient Refrigerators and Implied Consumer Discount Rates." Paper presented at the Second ACEEE Summer Study on Energy Efficiency in Buildings, Santa Cruz, CA, August 1982.

Ruderman H., M.D. Levine, and J.E. McMahon. "The Behavior of the Market for Energy Efficiency in Residential Appliances Including Heating and Cooling Equipment." LBL-15304, Lawrence Berkeley Laboratory, Berkeley, CA, September 1984.

U.S. Department of Energy, Conservation and Renewable Energy. "Consumer Products Efficiency Standards Engineering Analysis Document." DOE/CE-0030, March 1982a.

U.S. Department of Energy, Conservation and Renewable Energy. "Consumer Products Efficiency Standards Economic Analysis Document." DOE/CE-0029, March 1982b.

U.S. Department of Energy, Energy Information Administration. "State Energy Fuel Prices by Major Economic Sector from 1960 through 1977." July 1979.

U.S. Department of Energy, Energy Information Administration. "National Interim Energy Consumption Survey (NEICS)." 1980.

# Why Don't People Weatherize Their Homes?: An Ethnographic Solution[1]

*Richard R. Wilk*
*New Mexico State University*
*Harold Wilhite*
*Resource Policy Group, Oslo*

## INTRODUCTION

Most studies of energy use and conservation in this country over the last decade have assumed a relatively straightforward relationship between changes in the energy environment (i.e. policy, pricing regulations), and consumer behavior. The underlying assumption is that, given the proper incentives and adequate dissemination of the facts, consumers will react appropriately in a "rational" economic fashion.

Nevertheless, the study of consumer choice remains one of the least developed parts of economic theory (Haines 1972; Mason 1983). The relationship between supply, price, and demand for residential energy is complex and quirky. Even when details of the physical environment and the housing stock are well known, actual demand and use are difficult to predict with any accuracy (Penz and Yasky 1979). It is useless to merely lament the lack of

*Energy Efficiency: Perspectives on Individual Behavior*


---

[1] Research sponsored by the Universitywide Energy Research Group of the University of California.

"proper" response on the part of consumers or claim that cultural or social factors somehow get in the way of sensible action. Instead, we need more accurate predictive models of behavior, models that expand our notion of consumer rationality by accounting for social and cultural behavior factors.

In recent years, behavioral scientists have undertaken a number of empirical studies of consumer behavior that aim at a more complete theory of demand and consumer response. The most successful are those that model the processes of choice and try to determine which social and economic variables affect those choices (e.g. Gladwin 1980; Kempton and Montgomery 1983). A detailed knowledge of consumer choices is acquired through intensive, open-ended studies (often labeled "ethnographic"). Although ethnographic studies do not usually produce statistically valid results that can be freely generalized to the larger population of energy users, they do yield finely grained and detailed information that cannot be obtained through questionnaires, and they often provide unexpected insights and lead to productive new lines of inquiry. Most important, they place consumer choices within a wider context of other life decisions and link consumption to other processes and activities in society in general. Their results can later be tested in larger-scale statistical studies.

This chapter addresses energy-consuming behavior using such an ethnographic approach. Specifically, we examine the intriguing case of home weatherization: the weatherstripping of doors and windows and the caulking of joints and cracks. These are expand our notion of what really is "rational" to consumers.

Before we proceed to an analysis of consumer behavior with respect to weatherstripping and caulking, we should say a bit more about what weatherization is and what it does for a house. Its purpose is to slow down air exchange rates by reducing infiltration, and to thereby save on the cost of space heating and cooling. Estimates of the effectiveness of these measures in American houses vary considerably, because air infiltration is difficult to measure accurately. The amount of savings expected will depend on temperature differentials, pressure differentials (due to wind), the type and fit of existing windows and doors, mode of house construction, and the choice of weatherstripping and caulking methods (ASHRAE 1972).

In houses that have not been weatherstripped or caulked, air infiltration can be responsible for between 10 and 50 percent of heating load (Gibbons and Chandler 1981). Weatherstripping doors and windows alone can reduce infiltration losses by 33% or more (Luetzelschwab 1980), and caulking of cracks and sealing of other openings can push this figure above 66% (Blandy and Lamoureux 1980). Total savings on heating and cooling costs are likely to be between 10 and 20 percent for housing constructed before 1970.

A guide for consumers published by the U.S. Department of Housing and Urban Development in 1975 estimated the cost of caulk and weatherstripping installed by the homeowner at between $75 and $105 per house. For areas like northern California, annual savings were estimated between $30 and $75, with a payback period of between one and 3.5 years (HUD 1975). With continuing increases in fuel costs and conservation tax credits, these payback periods are even more favorable now than in 1975.

A number of other factors should combine with the short payback period to make weatherstripping and caulking attractive investments for homeowners. The measures often lead to an immediate increase in comfort (as drafts are reduced). The materials and technology are simple and can be installed without professional guidance or assistance. Furthermore, no radically new technology is being used, and weatherstrippping and caulking certainly cannot be considered "innovations" in the conventional sense of the term.

Given these incentives, what has been the response? Comprehensive and reliable statistics are not presently available; many surveys of household energy conservation do not include weatherstripping and caulking at all (Eichen and Tukel 1982). Available figures for the proportion of homes that have been caulked and weatherstripped vary widely, with some indications that the number of such homes has increased over time. Table 1 summarizes the results of several studies, which show that between 15 and 54 percent of surveyed households report some weatherization efforts, though in many cases the results fall far short of complete weatherstripping and caulking. It should be noted that most of these figures are for self-reported conservation measures, which tend to produce highly inflated percentages, as those questioned respond to the perceived expectations of the researchers (White et al. 1983). It is likely that actual performance of weatherstripping and caulking is considerably below these estimates.

In the light of these surveys, it is important to ask why more households have not followed these practices, and why other energy conservation measures that may have longer payback periods and

**Table 1. Proportion of Households That Have Weatherstripped in Different Studies**

| Survey | Date | Area | Proportion of Households Weatherstripped/ Caulked |
|---|---|---|---|
| Perlman and Warren | 1974 | USA | 17.0% |
| Cunningham and Lopreato | 1975-6 | Southwest | 41.3% |
| Olsen and Cluett | 1978 | Seattle | 54.0% |
| Jackson | 1978 | Alberta (Canada) | 14.9% |
| Jeppesen | 1982 | Michigan | 37.0% |
| Chatelain | 1981 | Utah (rural) | 33.0% |
| | 1983 | Utah (urban) | 35.1% |
| Feldman, Awad, and Williams | 1982 | S. California | 47.0% |

require higher investments are in some cases chosen over weatherization. Answers to these questions bear directly on the perplexing problem of the poor response of Americans to conservation incentives and programs, and the disappointingly small reductions in total household energy usage (Frieden and Baker 1983; Morrel 1981).

Our research in Santa Cruz County, California during 1982 and 1983 did not begin by focusing on weatherization. Rather, we were concerned with the processes of household energy decision making, asking how households made decisions to either invest in energy-conserving measures for their houses or to be more rigorous in their home energy management. Our study was based on open-ended ethnographic interviews, which allow informants to give their own explanations for their actions and permit the interviewers to expand on points of interest that arise (See Wilk and Wilhite 1983).

As we collected information on retrofit installations from people who had spent thousands of dollars on solar water heating, and on management decisions from people who spent their winters in multiple layers of sweaters, we became intrigued by the absence of any serious interest in weatherstripping and caulking, and the rarity with which such improvements were made. While household members were frequently enthusiastic and loquacious about their fireplaces, water heaters, skylights and thermopane glass, even those who had weatherstripped were reluctant to discuss their efforts and rarely volunteered information. Soon after the study began, we began to quiz people about weatherstripping, finding out the state of their knowledge, and eliciting a wide variety of explanations for why it was or was not a useful investment.

Given this puzzle—"Why is a perfectly valid conservation method virtually ignored by consumers?"—we chose to delve further into the matter. For though weatherstripping and caulking may be exceptional in their lack of "glamour", an investigation of the origin of this situation is likely to shed light on the converse situation—on why other measures do attract consumers, and are frequently adopted even when they do not provide immediate or even long-term economic returns.

Of our 60 sample households, 7 had purchased houses that were already weatherstripped. This means that weatherstripping was an economically viable addition to the house for 53 households. Of these, only 14 (about 26%) installed weatherstripping. Twenty one of the households that had not invested in weatherstripping had spent an average of about $1900 on measures such as solar water heaters, wood stoves, insulation and greenhouses. These other measures are more costly and have much longer payback periods than weatherstripping.

This behavior cannot be explained by a lack of knowledge. Every homeowner interviewed was aware of the costs of heat leakage as well as the benefits of weatherization, though homeowner's estimates of the amount that could be saved varied widely. Furthermore, those interviewed had often sought information about other conservation measures such as solar water heaters, while only a few had sought information on weatherization.

If the homeowners' purpose in installing energy conservation measures is to save money on fuel, weatherization should be much more common, and considered much more desirable, than it is in our sample. If consumers are aware of the benefits and can bear the costs (and nobody in our sample was too poor to afford weatherization), why don't they do what is best for themselves? Even when weatherization is *given* away to low income households, people often seem reluctant to have it done to their houses. Where do we look to attempt to solve this puzzle? We looked empirically at our detailed ethnographic data, and found that an answer requires an understanding of the cultural context of energy conservation. We find a key to be the question of the goals of consumers.

## INTERVIEW DATA

As mentioned above, information on weatherstripping was gathered as part of a wider ethnographic study of energy decision making in Santa Cruz County, California.[2] Weatherization was only one of many topics discussed, and we found the context in which weatherstripping was introduced to the conservation to be revealing. Each household was asked what energy conservation measures they had taken in the house, and of the 14 households that had either partially or completely weatherstripped their doors and windows, only 8 included weatherization in their listing. Of the 8 who did bring it up, 2 were responding to the question "What was the first thing you did when you moved into the house?" All of those who brought it up did so after discussing virtually every other measure that they had taken, usually after being prompted by the question "Are there any other energy conservation investments that you have made?" Evidently, when weatherstripping is done by a household, it is either considered to be of lesser significance than other measures, is not considered to be an energy conservation measure, or it is stigmatized in some way and people are reluctant to discuss it. This situation is different from responses about other energy investments. Solar panels and greenhouses are shown and discussed with pride, lower thermostat settings and extra sweaters are paraded as stoic adaptations, but weatherstripping is most often passed over and ignored.

Thus the 14 households that had purchased solar hot water heat-

[2] The sample consists of 60 middle class, Anglo-American households in Santa Cruz County, California, 30 of which had spent more than $50 on energy conservation measures for their houses, and 30 that had spent less than $50. The two groups, conserving and non-conserving, were matched according to the stage of the household developmental cycle (childless married couples, married couples with young children, with older children, and whose children had moved away, and non-family households). Interviews were about an hour in length and were conducted with as many household members present as possible. In each interview, the interviewer asked for an oral history of each energy decision and introduced for discussion several issues surrounding household decision making.

ing systems, and the 14 households that had bought wood stoves, were much more vocal about their purchases than the 9 households that had invested in weatherization. This situation is unusual because the payback period for an optimal solar installation in Santa Cruz County is more than ten years, with a similar interval for woodstoves if wood is purchased. Table 2 lists the actual conservation actions taken by the thirty households that had invested more than $50, demonstrating what a minor component of conservation strategy the weatherization process is. How can we understand this behavior? What lessons are there in these purchasing decisions?

## FOLK EXPLANATIONS

Ethnographic studies yield two levels of explanatory data. The first is the explanations offered by respondents for their actions in response to direct or indirect questioning. The second comprises synthetic explanations composed by ethnographers as interviews are dissected and analyzed. In this second stage the statements of the respondents are placed in a cultural context; deeper structures are sought behind the folk explanations. Thus the first level of explanation becomes data for the second level of analysis, though the folk explanation remains an important reference. Here we will briefly discuss what our respondents told us about weatherization before moving on to our own interpretation.

As our interviews progressed and we began to wonder why so few people were weatherizing their houses, we began to ask direct questions:

"Have you thought about weatherizing your doors and windows?"; "You say there's a draft under the door. Why haven't you weatherstripped it?"; "If your [RCS] audit said to weatherstrip and caulk, why did you put in the windows instead?". Phrased this way, we obtained several distinct types of responses.

A few people stated that weatherstripping was not an effective strategy for their houses, claiming that the structure was just too leaky, or that it just never got cold enough to make the effort worthwhile. More common were responses that acknowledged the cost effectiveness of weatherization while questioning the ramifications for health, a common concern in all discussions of home space heating.

It is a common perception, probably founded in the desire to reduce thermostat settings and energy use, that "fresh air" and

**Table 2. All Investments in Energy Conservation for a Sample of 30 Santa Cruz Households, Numbered from One to Thirty.**

| Investments | Households (by number) | | | | | | | | | | | | | | |
|---|---|---|---|---|---|---|---|---|---|---|---|---|---|---|---|
| | 1 | 2 | 3 | 4 | 5 | 6 | 7 | 8 | 9 | 10 | 11 | 12 | 13 | 14 | 15 |
| Weatherstrip/Caulk | X | | | X | X | | | | | | X | X | | | |
| Solar Water | | | X | | X | | | X | X | | X | X | | | X |
| Insulation | X | X | X | X | | X | | | X | X | X | X | | X | |
| Wood Stove | X | | | | | X | X | X | X | X | | | X | X | |
| Heavy Curtains | | | X | | | | | | | | | | X | | |
| Double-Pane Glass | | X | | X | | | | | X | X | | | X | | |
| Water Heater Timer | | | | | | | | | | X | | | | | |
| Water Heater Jacket | X | X | | | X | X | X | X | | X | X | X | X | | |
| Fireplace Gadgets | X | | | | | | X | X | | | | X | X | | X |
| Flourescents | | | | X | | | | X | | | | | | | X |
| Greenhouse | | | | | | X | | X | | | | | | | |
| Heat Exchanger | | | | | | | | | | | | | | | |
| Pipe Insulation | | | | X | | | | | | | | | X | | |
| Trombe Wall | | | | | | | | | | | | | | | |

**Table 2. (Continued).**

| Investments | Households (by number) | | | | | | | | | | | | | | |
|---|---|---|---|---|---|---|---|---|---|---|---|---|---|---|---|
| | 16 | 17 | 18 | 19 | 20 | 21 | 22 | 23 | 24 | 25 | 26 | 27 | 28 | 29 | 30 | Total |
| Weatherstrip/Caulk | | | X | | | | | | X | X | | | | X | | 9 |
| Solar Water | | X | X | | | X | | | | X | | | X | X | X | 14 |
| Insulation | | | X | X | X | X | | | X | X | | | | X | X | 18 |
| Wood Stove | X | | | X | | | X | | X | | | | X | X | | 14 |
| Heavy Curtains | | | | | | | | | | | | | | | | 2 |
| Double-Pane Glass | | | X | | | | | | X | | | | | | | 7 |
| Water Heater Timer | | | | | | | | | | X | | | | | | 2 |
| Water Heater Jacket | | | X | X | | | X | X | | | | X | X | | | 16 |
| Fireplace Gadgets | X | X | | | | | | | | | X | X | | | | 10 |
| Flourescents | | | | | | | | | | X | X | | | | | 5 |
| Greenhouse | | | X | | | | | | | | X | | | | | 4 |
| Heat Exchanger | | | | | | | | | | | | X | | | | 1 |
| Pipe Insulation | | | | | | | | X | | | | | | | | 3 |
| Trombe Wall | | | | | | | X | | | | | | | | | 1 |

lower temperatures are more healthy than "stuffy warmth." Coupled with media reports of the dangers of over-sealing and indoor air pollution in new homes, a folk theory has developed that posits health benefits for high air-infiltration rates (and that confuses weatherization with insulation as a source of danger).

While the factual basis for such fears is minimal (there is little possibility that the infiltration rate after retrofitting can or will be reduced to a dangerous level) the folk model has tangible effects. Often, however, the issue of health seems to be more a convenient rationalization for inaction than a part of a concrete plan of action for promoting family health.

Several respondents said that, while they knew weatherization would benefit their houses and save money, they did not have the technical competence to do the job themselves (despite the fact that some had taken on major home improvement projects). There seems to be confusion in the marketplace about the best kinds of weatherstripping, and there is a good deal of folk-knowledge in circulation about the problems of some kinds of installations. Some people mentioned friends who had bad experiences with weatherstripping materials that had lasted only a short time, or expensive door channeling that never fit correctly and didn't keep drafts out. In a couple of instances, people had tried to weatherstrip without adequate preparation, on a trial basis, and the failure of their initial effort served to rationalize further inaction. Several of those who had installed weatherstripping volunteered "I don't feel that I have put it on right."

Rather than declaiming their lack of knowledge or expertise, several people responded by telling us that they were planning to weatherize, but just hadn't gotten around to it yet. On further probing we found that many had been "planning" to weatherize for years; it just never seemed to get to the top of the priority list.

> *Husband:* We should have the doors weatherstripped.
> *Interviewer:* Why do you think you haven't done it?
> *Husband:* I don't know, I just didn't get to it. But that's what should be done.
> *Interviewer:* How long have you been thinking about it?
> *Husband:* Quite a while.
> *Wife:* For years!

One man admitted that he had had all the materials and tools sitting in a drawer for three years! These people showed awareness of the benefits, understood how little money and effort was involved, and felt a bit guilty and defensive because they had never done it. When pressed, they were able to give a number of reasons (rationalizations) for "putting it off and putting it off." One man said that "it is one of those things that you don't have time to do until sometime, and then you go sailing."

One of the most interesting, and common responses was that the job of weatherizing is a "dirty" one, which is "unpleasant", contrasting with jobs like painting and remodeling, which produce aesthetically pleasing and rewarding results.

Related to this "non-glamorous" perception of weatherstripping is that it is "small", "invisible", and a sort of "plugging up the cracks". In the following excerpt the wife thinks her husband has not weatherstripped because the job is "small", while the husband uses a common and very effective form of rationalization that we call "linking".

*Interviewer:* What is it about weatherstripping?

*Wife:* Maybe because it seems small, I don't know.

*Husband:* I tell you what it is, because if I do weatherstripping I have to paint the windows. It just seems a logical way to do it.

*Wife (with irony in her voice):* Is that why you don't do it? I just thought other things were more important.

In the above excerpt, the husband has linked the unpleasant job of weatherstripping to another job, painting the windows. Sometimes linkage produces whole chains of obstacles. These barriers may flow from real physical constraints to activities, but most often linking reflects a conscious or unconscious priority list. Commonly a somewhat vague list of chores and jobs is kept; linkage provides the rationale for keeping undesirable jobs like weatherstrippping down at the bottom of the list virtually forever.

These folk models on the part of non-weatherstrippers contrast sharply with the statements of those who did weatherstrip and caulk. Here we find statements about how dramatic the improvement was, how obtrusive the cold drafts were beforehand and how much more comfortable things were afterwards. Sometimes we detected a note of defensiveness about these answers, as if people were having to account for actions which might otherwise be misinterpreted. Nobody said they liked the job, even when they praised the results. Some did remark that it was not as difficult or dirty as they expected, which raises the question of why they expected it to be so bad in the first place. We also found that people spoke of their weatherization at the same time they were listing the things they had done to "fix up" their houses after first moving in, rather than discussing it along with the "home improvements" they undertook after they had occupied the house for some time. This fact proves to be an important key in interpreting the inactions for others, but is a second-level interpretation of the data and will be treated below.

## ANALYSIS AND DISCUSSION

We have other data on the households in our sample besides their statements about weatherization. Though all interviewed households fall into a fairly narrow range of income, they vary widely in the number of years they have occupied their houses, and in the stage of their family development. Table 3 breaks down those households that have weatherstripped by stage in the family developmental cycle. Twelve of the fourteen are from the earlier phases (unmarried couple, married couple with no children, and married couple with young children). Nine of the fourteen bought their house as a "fixer-upper", with the foreknowledge that they would have to put a lot of work into it to bring it up to their standard. Since younger married couples are generally the ones that are prepared to buy a house that is in a sense "incomplete" and to do much of the work themselves, these two bits of information are mutually reinforcing.

We can also look at the sequence of different kinds of retrofits. Analysis of the order in which our sample households undertook energy conservation investments shows that when a house is weatherstripped, it is usually done shortly after a home is occupied, long before other energy conservation measures are considered or purchased. If weatherstripping is not done during the initial two years after the home is occupied, it is unlikely that it will be done at a later date. If it is done later, it is almost always done in conjunction with a major remodeling project, rather than by itself.

These data on the timing of weatherization efforts contrast strongly with those for home insulation, which tends to be undertaken by more mature households, later in the developmental cycle. Insulation is rarely installed immediately after a house is occupied, but rather comes after the household has settled in, often when total income is higher, but disposable income is lower.

The major facts that need explaining are: first, that few people ever weatherize at all, despite the manifest economic advantages

**Table 3. Number of Households Weatherstripping by Phase of the Family Development Cycle.**

| | Group 1 (total investment > $50) | | Group 2 (total investment < $50) | |
|---|---|---|---|---|
| Married Couple, no children | 1 | N=3 | 2 | N=3 total 3/6 |
| Married Couple, children < 12 yr. | 4 | N=10 | 2 | N=10 total 6/20 |
| Married Couple, teenage children | 0 | N=3 | 1 | N=3 total 1/6 |
| Married Couple, children gone | 1 | N=6 | 0 | N=6 total 1/12 |
| Non-nuclear family household | 3 (all young) | N=8 | 0 | N=8 total 3/8 |
| Total | 9 | N=30 | 5 | N=30 |

and the potential for increased comfort; and second, that those who do weatherize are a particular sub-group comprising younger households, especially those who have just moved into "fixer-upper" houses. In light of people's own statements about weatherizing, we can summarize some perceptions that must be taken into account. These are that weatherstripping is a "dirty" job, that it is relatively invisible, that it does little to improve the appearance or value of the home, and that it is not quite a big or important enough job to make it worth bringing in a specialist or contractor.

We find that our conversations with homeowners provide an ethnosemantic framework that shows how all of these observations fit together. An ethnosemantic analysis delineates the semantic "fields" or "domains" that structure thought and action (see Bruner et al., 1956; and Frake 1972). In other words, the ways that people classify objects and actions have a direct affect on behavior. Those things that are ambiguous and do not fit comfortably into the classificatory structure, tend to be avoided, shunned, or even tabooed in extreme cases (Douglas 1966).

Given this background, it is possible to demonstrate that weatherization is a semantically ambiguous task. Weatherization is certainly not classified by our respondents as a form of maintenance, though in fact most kinds of weatherstripping and caulk must be repaired, checked and even replaced regularly. Nor does weatherization fall into the category of a home improvement, for these activities are generally highly visible additions to the home that substantially increase the actual monetary value of the structure.

The category of "repair" is an uneasy place for weatherization as well, for people tend to see repairs as actual fixing of something that is broken. A leaky door still functions as a door, it is hardly "broken" and in need of repair. In making up their lists of "things to do around the house", urgent repairs usually sit at the top, with highly desirable improvements just below. Even when weatherization is acknowledged to be a repair or an improvement, it always remains low on the list. It remains a weak and ambiguous member of the categories except under exceptional circumstances.

Moving into a new house provides just those special circumstances. At this time, especially if the house is a "fixer upper", everything that can possibly be repaired is taken care of in one long extended effort that may last for some time. Many minor repairs that might be left for years by someone living in a house will be undertaken when someone moves into a house for the first time. But once the "grace period" is over, once the house is seen as "complete" in the eyes of the household, meaning that it is no longer in need of repair, a device must actually break to initiate repair. After this grace period, anything not broken must be classified as an "improvement" to the house before it is undertaken. Otherwise, a kind of semantic embarrassment would occur; the household would have to admit that what they had believed a complete functioning house had actually been in need of repair (weatherization) all along. No wonder people react with annoyance to salespeople who tell them that their house urgently needs weatherization; the implication is that there is something wrong with their house that they haven't had the sense to see. Somehow their integrity is impugned.

Middle-class Americans tend to identify closely with their houses; they focus much of their emotional life on home and family. An attack on the integrity of the house is easily translated emotionally into an oblique and indirect attack on the solidarity of the family. Repairs and repairmen are seen as intrusive, while home improvement is much more positive, interesting and constructive. The most prominent theme running through our interview data is that energy conservation retrofits are much more likely to be pursued when they fall into the category of home improvement; in fact, we assert the American homeowner considers energy conservation to be a small sub-category of home improvement. It often comes much lower on the priority list than gardening and landscaping.

It is possible to ask at this juncture why people do not simply categorize weatherization as a form of home improvement, something that will make their house better, rather than something to fix what is wrong or broken. Those few people we spoke to who had undertaken weatherization long after moving into their homes focused mainly on the increase in comfort they had gained from eliminating drafts of cold air. But they were a bit embarrassed about it nonetheless, and our impression was that they wanted to avoid censure for being "stingy" or "tight". Overzealousness in saving money is not a virtue in middle-class life.

After all, weatherization really is stuffing up the cracks and is not exactly in the same league as a new bedroom, a greenhouse, or a solar panel. At its best, weatherization is a kind of negative improvement - it decreases loss rather than increasing gain. Woodstoves or solar heaters actually produce energy, giving the impression of "something for nothing", while insulation and weatherization suffer because they produce nothing.[3]

Other improvements are visible to neighbors, serving social ends. Weatherstripping on the other hand, is not glamorous; "...you're not exactly about to show off your weatherstripping to your neighbors...they can't even see it if you do it right." Ideally, an improvement should give some kind of feedback to tell the owners that it is functioning correctly, serving as a focus for justification and pride. The gurgle of water in hot water pipes (even when it is a sign of air bubbles in the system) is seen as rewarding and is pointed out to guests, as is the heat from a woodstove, even when hands are burned and the room smells of smoke. Weatherization offers little feedback and few opportunities for bragging. Sometimes it is even hard to tell if it is working correctly; air leaks are invisible.

---

[3] While energy specialists often consider conservation to be a form of energy "production", homeowners draw a sharp distinction between conservation and production.

## CONCLUSIONS

We began this discussion by challenging a narrow economic approach to understanding consumer choices. We have now incorporated a wide range of cultural and psychological factors that must be taken into account in order to make sense of consumer reticence when it comes to weatherization. The behavior we have been studying is indeed rational, regularized goal-seeking that can be studied systematically and formally. The crucial issue however, when dealing with as culturally important an arena as the home, is deciding which goals are being pursued. Certainly weatherstripping is a highly rational action if one's foremost goal is saving money on heating bills. But the homeowners we have talked to have expressed, and pursued, many other goals besides cutting their utility expenditures.

We do not want to imply that those in our sample are uninterested in saving money, for nobody told us that they didn't care how much money they spent on heating. Everyone showed concern for rising bills, and a small minority explicitly mentioned cash saving as their major reason for pursuing energy conservation methods of all kinds. But this was a minority position, for most of our respondents had many other goals when modifying their home. Unless a device, installation, or improvement meets some of these other goals in addition to that of saving money, it is unlikely to be adopted.

Just what are these other goals? We have already touched upon important goals—that of improving the home's aesthetic and monetary value in the eyes of occupants and neighbors and that of increasing the comfort of the home environment. One constant and salient theme in interviews with those who have made investments in solar water heater and woodstoves is the desire for independence. In brief, many Americans feel that their home is their last refuge of privacy and individuality, and they tend to resent intrusion. One of the most obvious kinds of intrusion is the utility company—large, impersonal, and characterized as rapacious. Independence can be augmented by reducing dependence on utilities, and this reason was constantly mentioned as a justification for energy conservation -- especially those kinds of conservation that entail actual energy production. While the independence thus obtained is symbolic and metaphorical, it is important nonetheless.

Within the general category of values entailing independence we find a minority position motivated simply by hatred of utility companies more than by a desire for symbolic independence. A similar minority ideological value is found among those whose interest in energy conservation stems from a concern with global depletion of non-renewable resources. Again, the concern is with reducing the amount of energy purchased from the utilities, but more through substitution than through "stuffing up the cracks." The desire to do something about pressing global problems requires some kind of regular satisfaction, some tangible and visible proof, and simply reducing air infiltration is insignificant in this respect.

Of all of the goals we have mentioned, actual monetary saving on energy bills is the only one that is directly addressed by weatherization; the rest are best satisfied by other measures. It is quite ironic then, that it is just because the benefits of weatherization are so clearly economic that it is so unpopular! This is why it lacks "glamour" and appeal on the marketplace. It is also why so many of our respondents are willing to spend thousands of dollars on solar water heaters that can never pay for themselves, while they are unwilling to take an afternoon to caulk cracks and save hundreds of dollars in the first heating season. As specialists we may laugh at such behavior, but in the interests of further energy conservation in the country it behooves us to learn to work with the existing set of cultural values rather than to challenge them.

**References**

American Society of Heating, Refrigeration, and Air-Conditioning Engineers (ASHRAE). *Handbook of Fundamentals.* New York: ASHRAE, Inc., 1972.

Blandy, T. and D. Lamoureux. *All Through the House: A Guide to Home Weatherization.* New York: McGraw Hill, 1980.

Bruner, J., J. Goodnow, and G. Austin. *A Study of Thinking.* New York: John Wiley and Sons, 1956.

Douglas, M. *Purity and Danger.* London: Routledge & Kegan Paul, 1966.

Eichen, M. and G. Tukel. "Energy Use and Conservation in the Residential Sector." *Energy Policy,* March 1982.

Frake, C. "The Ethnographic Study of Cognitive Systems." In *Culture and Cognition*, pp.191-225. San Francisco: Chandler, 1972.

Frieden, B. and K. Baker. "The Market Needs Help: The Disappointing Record of Home Energy Conservation." *Journal of Policy Analysis and Management*, March 1983.

Gibbons, J. and W. Chandler. *Energy: The Conservation Revolution.* New York: Plenum, 1981.

Gladwin, C. "Cognitive Strategies and Adoption Decisions." In *Indigenous Knowledge Systems and Development*, ed. D. Brokensha, D. Warren, and O. Weiner. Washington, D.C.: University Press of America, 1980.

Haines, G. "Overview of Economic Models of Human Behavior." In *Consumer Behavior: Theoretical Sources,* ed. Ward and Robertson, pp. 276-300. Englewood Cliffs, NJ: Prentice-Hall, 1972.

U.S. Department of Housing and Urban Development, Office of Policy Development and Research, Division of Energy, Building Technology, and Standards. *In the Bank...or Up the Chimney?* Washington, DC: U.S. Department of Housing and Urban Development, 1975.

Kempton, W. and L. Montgomery. "Folk Quantification of Energy." *Energy* 7:817-827, 1983.

Luetzelschwab, J. *Household Energy Use and Conservation.* Chicago, IL: Nelson Hall, 1980.

Mason, R.S. "The Economic Theory of Conspicuous Consumption." *International Journal of Social Economics* 10(3):3-17, 1983.

Morrel, D. "Energy Conservation and Public Policy: If It's Such a Good Idea, Why Don't We do More of It?" *Journal of Social Issues* 37:8-30, 1981.

Penz, A. and Y. Yasky. "Uncertainties in Predicting Energy Consumption in Houses." *Energy Systems and Policy* 3(3):243-269, 1979.

White, L., D. Archer, E. Aronsen, L. Condelli, B. Curbow, B. McLeod, T. Pettigrew, and S. Yates. "A Meta-Evaluation of the Energy Conservation Research of California's Utilities". Universitywide Energy Research Group, Paper 107. Berkeley: University of California, 1983.

Wilk, R. and H. Wilhite. "Household Energy Conservation Decision-Making in Santa Cruz County, California". Universitywide Energy Research Group, Paper 105. Berkeley: University of California, 1983.

# Energy Conservation and Public Policy: The Mediation of Individual Behavior

*Dane Archer, Thomas Pettigrew, Mark Costanzo, Bonita Iritani, Iain Walker, and Lawrence White*
*University of California, Santa Cruz*

## INTRODUCTION

Energy conservation constitutes an important enigma for public policy. Broad analyses of the persistent "energy crisis" have concluded that conservation is indispensable to any solution (Stobaugh and Yergin 1979), and studies of public opinion indicate pervasive concern about energy and widespread support for conservation (Olsen 1981). Despite these auspicious premises, energy conservation remains an area in which concrete accomplishments have been somewhat disappointing.

Actual levels of energy consumption continue to defy simple formulations based on architecture (Socolow 1978; Seligman et al. 1978), suggesting that the unique qualities of individual consumers are as important as the inherent energy efficiency of a given building. The factors that govern individual consumption levels remain largely unknown. Surveys commonly find that energy behavior is not readily explained by individual attitudes toward energy and conservation (Anderson and Lipsey 1978) and, partly as a result, public policies and experimental programs designed to further

*Energy Efficiency: Perspectives on Individual Behavior*

energy conservation have reflected a confusing theoretical patchwork of approaches (Stern and Gardner 1981; Morrell 1981).

In the context of this fundamental uncertainty about the antecedents and "causes" of energy-conserving behavior, it is not surprising that even the most costly and ambitious energy conservation programs have shown modest results (Seligman and Hutton 1981; Archer et al. 1983; Condelli et al. 1983). Some of the most extensive of these programs have been conducted by large utility companies. Reviews of these programs have criticized design features that undermine the effectiveness of a program or make its evaluation impossible (White et al. 1983).

## THE "ATTITUDE" AND "RATIONAL" THEORETICAL MODELS

In addition to important methodological defects, many of the largest energy conservation programs have relied, explicitly or not, on two vague theories concerning conservation behavior. The first might be called the *attitude model* of conservation behavior—the assumption that favorable attitudes lead to conservation behavior. This model further assumes that making people's attitudes more favorable will make them more likely to practice conservation.

The second vague theory might be called the *rational model* of conservation behavior—the assumption that people will perform conservation behaviors if these behaviors are economically advantageous and, further, that increasing fiscal incentives will make these behaviors more probable. According to the rational model, decision-making results from an informed assessment of costs and benefits.

Although rarely stated in precisely this form, these two models give justification for many energy conservation programs. Influenced by the attitude model, utility companies have pursued extensive and costly advertising programs to induce consumers to conserve (Archer et al. 1983; Condelli et al. 1983). Influenced by the rational model, the federal government, individual states, and specific utility companies have enacted tax credits, product rebates, and low or zero-interest loans to promote conservation and the acquisition of conservation devices (e.g., California Energy Commission 1981).

These two theoretical premises are intuitively reasonable, and it is scarcely surprising that they have provided the foundation for so much public policy and so many conservation programs. Like all theories, of course, these two pervasive models imply testable propositions about real people and their behavior. In this paper, aspects of these two theoretical models are tested using new survey data. The emphasis in these data is on problematic links between policy efforts to encourage conservation, on the one hand, and the acts of individual consumers on the other. On the basis of this analysis, more pragmatic and theoretically more promising approaches to energy conservation policy are suggested.

**Attitudes and the Limits of the Attitude Model**

As a test of whether these conceptions implicit in public policies correspond to actual behavior, we conducted a telephone survey of a probability sample of households in Santa Cruz County, California between November 1981 and March 1982. The random-digit-dialing (RDD) method was used to ensure reaching unlisted as well as listed residential phones (Groves and Kahn 1979; Waksberg 1978). We completed 642 interviews for a satisfactory response rate of 73.2%, with most interviews requiring between 10 and 15 minutes. A thoroughly bilingual interviewer, using a Spanish-language version of the survey, questioned all respondents who wished to be interviewed in Spanish.

On all grounds, this coastal California county is an optimal site for an energy conservation survey. The county includes a major university campus, a community college, and other "progressive" institutions. Environmental concerns appear to be strong in this county, and the abundant sunshine combined with cool maritime temperatures makes alternative fuels such as solar energy attractive.

Many of our respondents have a favorable attitude towards energy conservation, and most believe that the energy situation is a serious crisis. When asked, "How serious do you think this country's energy situation is right now?", 43.1% of our respondents report that it is "extremely serious" and another 41.9% think it is "somewhat serious." When asked whether they feel the energy situation will improve, remain the same, or become worse in the next ten years, more than half (54.5%) believe it will become worse. These two survey items are significantly related to one another (chi-square = 36.33, $p < 0.001$). Those who think the energy situation is serious also say it will worsen. This grim expectation characterizes roughly half of our respondents; 49.2% indicate they believe the energy situation is presently serious and will become worse.

**Table 1. Attitudes Toward the Energy Situation**

| Characteristic (n) | | Percent who said the energy situation is serious* | Percent who said the energy situation will become worse |
|---|---|---|---|
| **Education** | | | |
| H.S. or less | (291) | 79.4‡ | 49.8 |
| College | (276) | 88.8 | 58.0 |
| Grad School | (75) | 93.3 | 60.0 |
| **Age** | | | |
| 18–24 | (94) | 84.0‡ | 59.6† |
| 25–34 | (172) | 89.5 | 62.8 |
| 35–44 | (120) | 92.5 | 57.5 |
| 45–54 | (52) | 84.6 | 44.2 |
| 55–64 | (79) | 83.5 | 54.4 |
| Over 65 | (95) | 69.5 | 37.9 |
| **Gender** | | | |
| Female | (339) | 89.4‡ | 56.3 |
| Male | (205) | 77.6 | 54.6 |
| **Home Ownership** | | | |
| Rent | (254) | 87.0 | 61.0** |
| Own | (388) | 83.8 | 50.3 |
| **Family Income** | | | |
| Under $15,000 | (318) | 82.7 | 54.1 |
| $15,000–30,000 | (192) | 86.5 | 55.2 |
| Over $30,000 | (132) | 88.6 | 54.5 |

* Includes those who responded "somewhat serious" and "extremely serious."

** Chi-square significant, $p < 0.05$;

† $p < 0.01$

‡ $p < 0.001$.

The somber outlook found in our data is comparable to nationwide survey results. In a review of six studies, for example, Olsen (1981) discovered that between 40 and 60 percent of survey respondents believe that we are faced with an energy crisis that is both serious and chronic. In our survey data, this belief varies somewhat by demographic and other background variables, as shown in Table 1. As the figures in this table indicate, the perceived seriousness of the energy crisis is related to three variables: education, age, and gender. In addition, expectations for the future are related to age and home ownership.

At face value, these attitudes seem auspicious for energy conservation. Given these high levels of concern, one might well expect actual conservation behaviors to be widespread. The attitude model of conservation behavior implies a direct relationship between attitudes about the energy situation and actual conservation behavior. This model assumes that people conserve energy more fervently if they perceive that the "energy crisis" is becoming more serious. This theoretical model seems eminently plausible, and many energy policies and programs have been predicated on improving attitudes toward energy conservation (Archer et al. 1983; White et al. 1983; Condelli, et al. 1983).

Despite the apparent reasonableness of the attitude model, the research literature from social psychology suggests that the attitude-behavior link is rarely consistent, direct, or very strong (Ajzen and Fishbein 1977; McGuire 1968; Olsen 1981). Consistent with these somewhat counter-intuitive findings, our survey data contain no evidence of an impressive relationship between energy-related attitudes and behavior patterns. Individuals more concerned about the energy crisis or more likely to believe that the situation would become worse did not differ in general from other respondents in terms of their energy conservation behaviors. This attitude-behavior disjunction has obvious significance, since *behavior* patterns (and not attitudes) are the energy policy goals of consequence.

As a test for an attitude-behavior relationship in our energy data, analyses of variance were performed separately for the two attitude variables, each coded in three levels: (1) perceived seriousness of the current energy situation (not serious, somewhat serious, extremely serious); and (2) belief in the likely future of the energy situation (will improve, remain the same, will get worse). Each three-level attitude variable was combined with home ownership (owner, renter) to form two 3 x 2 factorial designs. Using each of these two 3 x 2 designs, four ANOVAs were run for the following four behavioral measures (described later in this paper): (1) the respondent's self-reported effort put into energy conservation; (2) Guttman scale of four self-reported conservation habits (closing drapes, conserving hot water, turning off lights, and recycling); (3) a

Guttman scale of four self-reported, one-shot, device-installing behaviors (installing insulation, weatherstripping, low-flow shower heads, and hot-water heater blankets); and (4) the total number of pieces of solar equipment installed in the home.

A total of eight ANOVAs were run, with a sample size of over 600. Despite the considerable statistical power and also the inflated alpha values in this analysis, only two significant effects (both weak in absolute size) were found for the attitude variables. Respondents who felt that the energy situation was extremely serious self-reported greater effort toward energy conservation ($F=3.47$, $df=2,636$ and $p < 0.05$). In addition, respondents who said they thought the energy situation would improve reported higher levels of conservation habits ($F=5.20$, $df=2,636$ and $p < 0.01$).

Even though eight ANOVAs were performed with this large sample, no other attitude differences even approached significance. In addition, there are reasons to be circumspect in assessing the two modest attitude differences that were found. For example, questions about self-reported habits tend to be value-laden and therefore invite bias (e.g., "Do you conserve hot water while showering?"). Finally, it should be noted that the practical significance of these relatively minor conservation habits seems doubtful. Evaluations of the relative importance of different conservation strategies consistently find that such habits have little effect on energy consumption and, in addition, are difficult to sustain (Stern and Gardner 1981).

As a second test for a possible attitude-behavior relationship, respondents were asked an open-ended question about the *reasons* for their expectations that the energy situation would improve, worsen, or remain the same. Answers were coded into several categories ("oil companies will find more energy", "coal or nuclear will become more important", "individuals will conserve", etc.). Again, there is little evidence for the attitude model. People who cited conservation as the most important factor in the energy future were, in fact, no more likely to practice it. These respondents did not differ from others on *any* of our conservation measures.

In summary, there appear to be no important relationships between respondents' attitudes toward the energy crisis and their behavior patterns. This lack of correlation raises serious doubts about the efficacy of energy conservation programs and policies predicated on the attitude model.

**Public Understanding and the Limits of the Rational Model**

Public understanding is a key element in rational theories of energy conservation. This theoretical model assumes that people need an *awareness of* and an *understanding about* the conservation incentives to determine whether a given device or change is cost-effective for them. The attitude model also implies prior understanding, since this model assumes that even if individuals are favorably disposed to practice conservation, they need to be *aware* of conservation programs in order to take advantage of them. Implicit in both theories, therefore, is the notion that high levels of public awareness and understanding are indispensable to widespread conservation. These theories also posit several other variables (favorable attitude, substantial personal energy costs, disposable income, etc.), but both theories regard awareness and understanding as *necessary* for successful conservation.

Our survey data indicate that this apparently reasonable assumption also contains serious flaws. Consider first the apparent poverty of public understanding. Even in an energy-conscious area in which attitudes toward conservation are highly favorable, levels of public understanding remain problematic. Table 2 presents data on public awareness and understanding of four conservation programs. Data are given on both "claimed awareness" (the respondents' stated familiarity with a given conservation program) and "accurate information" (the respondents' ability to provide concrete information about this program). This difference is similar to the distinction in marketing research between "claimed recall" and "proven recall"—in the latter case, the respondent is asked to provide some minimal evidence that the claimed awareness is genuine. In coding the solar tax credit of 55%, an answer between 45% and 65% was accepted as accurate.

The evidence suggests that the difference between claimed awareness and accurate information is substantial. Between roughly one-half and three-quarters of all respondents *claimed* familiarity with four conservation programs: (1) *peak load* usage and efforts to encourage use of electric appliances in nonpeak hours, (2) *home audits* of household energy consumption and needed conservation improvements, (3) *graduated rate structure* of utility charges to encourage conservation, and (4) *solar credits* that returned up to 55% of the cost of solar devices to the consumer in the form of tax credits.

As shown in Table 2, however, accurate or proven information

**Table 2. Public Understanding of Energy Conservation Programs: Claimed Awareness vs. Accurate Information***

| Awareness vs. Accurate Information | Program | | | |
|---|---|---|---|---|
| | Peak Load | Home Audit | Graduated Rate | Solar Credit |
| **Percent "Aware"** | 78.1 | 56.7 | 44.1 | 72.0 |
| Rent | 79.8 | 43.8 | 29.7 | 67.5 |
| Own | 77.3 | 64.8 | 53.8 | 75.0 |
| High School | 72.7 | 51.0 | 36.4 | 55.1 |
| College | 82.4 | 61.1 | 47.3 | 85.5 |
| Grad. School | 86.3 | 71.1 | 67.1 | 94.7 |
| Under $15,000 | 71.9 | 50.9 | 32.4 | 63.8 |
| $15,000–$30,000 | 81.5 | 58.9 | 47.9 | 80.2 |
| Over $30,000 | 89.1 | 67.4 | 65.6 | 86.4 |
| **Percent Accurate** | 41.3 | 1.4 | 14.3 | 13.2 |
| Rent | 42.2 | 0.0 | 9.6 | 10.0 |
| Own | 40.7 | 2.3 | 17.6 | 15.5 |
| High School | 38.8 | 0.0 | 11.0 | 6.8 |
| College | 42.8 | 1.8 | 16.4 | 17.0 |
| Grad. School | 47.4 | 5.3 | 21.1 | 23.7 |
| Under $15,000 | 39.8 | 0.5 | 7.7 | 5.9 |
| $15,000–$30,000 | 47.4 | 1.6 | 19.3 | 15.6 |
| Over 30,000 | 42.4 | 2.3 | 23.5 | 24.2 |

* Accuracy levels are generous. Respondents were given credit for an item if they were even partly correct—e.g., for the 55% solar tax credit, estimates between 45% and 65% were counted as correct.

about these programs was rare and, in some cases, negligible. In the case of three conservation programs, fewer than one respondent in eight understood the nature of the program, even using generous criteria for accurate understanding. This contrast in Table 2 demonstrates that studies of public understanding of policies and programs must include measures that tap *genuine* information levels, not merely claimed information levels.

In part, this apparent contradiction may reflect the well-known problem of "social desirability" (Crowne and Marlowe 1964; Rosenthal and Rosnow 1969)—in this case, the understandable desire of respondents to appear well-informed. This interpretation is supported by the relationship between awareness and background variables. Claimed awareness is more strongly related to the background variables in Table 2 (Median $c$ = 0.018) than is accurate awareness (Median $c$ = 0.12). This result suggests that higher status respondents may have felt, more than lower status people, a need to appear well-informed.

The contrast between levels of claimed awareness and proven awareness is not only interesting but consequential. Studies using only measures of *claimed* familiarity run the clear danger of greatly over-estimating the degree to which a policy has been successfully communicated. Since such measures are in common use in research on energy policy and energy conservation (Archer et al. 1983; Condelli et al. 1983; White et al. 1983), the evidence in this table has obvious implications for evaluation research.

The principal implication of these data for energy policy concerns the apparent potential for successful conservation programs. Our data suggest that accurate public awareness and understanding is a link that is largely missing, even if it is as vital a link as these theories imply. Coming on the heels of extensive utility advertising, and obtained in a region characterized by strongly pro-conservation attitudes (California), these survey data reflect surprisingly scant evidence of widespread public understanding of conservation programs.

If public awareness of conservation programs is indispensable to effective conservation, as attitude theory and rational theory both seem to imply, the evidence presented in Table 2 is discouraging. These theories assume that accurate understanding is directly tied to conservation behavior, and that low levels of the former necessarily are associated with low levels of the latter. This formulation implies other, more dynamic conclusions as well—for example, that *increasing* public understanding will also *increase* conservation behaviors. Both the attitude and rational theories assume that a link between understanding and behavior exists and, further, that this relationship is causal rather than merely associative.

Judging from our survey data, the link between understanding and conservation behavior is more tenuous and complicated than this simple formulation implies. In the case of several major conservation behaviors, there is little evidence that an understanding of conservation incentives and policies is a necessary condition, or even particularly important. For example, if one examines the decision to purchase four types of major solar equipment (a solarium, solar hot water heater, solar space heater, solar pool or spa heater), the role played by accurate understanding is far from clear. For each of these four major solar purchases, ownership of a solar device was only weakly related to accurate understanding of the solar tax credit (median *phi* = 0.05), and the relationship was in fact negative in one instance—solarium owners knew less about the solar tax credit than non-owners.

Accurate understanding does explain more of the variation if one eliminates renters from the analysis. For this smaller sub-sample, 27.9% of the home owners who have a major solar device understand the solar tax credit, while 13.9% of the home owners who do not have a major solar device understand the solar credit. This relationship is statistically significant (chi square = 4.71, p = 0.03) but weak (phi = 0.12). In addition, this finding shows that approximately three-quarters of the homeowners with a major solar device did *not* have even a crudely accurate understanding of the solar tax credit.

This result appears to contradict a *central* tenet of the rational model: that people determine whether important conservation devices are cost effective, and that they make this rational calculation by weighing relevant information about costs. Our data suggest that information indispensable to even gross cost calculations was, in fact, absent. It is possible, of course, that respondents forgot important details of conservation programs following their major purchase, or that a member of the household other than the respondent made the decision to invest in costly solar equipment. At the very least, however, this finding raises serious questions about the relationship between policy awareness and conservation behavior.

Note that this test of the rational model errs, if anything, in favor of the model. For one thing, far from assuming full and sophisticated information, we have coded respondent answers using extremely lenient criteria for accuracy. In addition, we have made no effort to correct or subtract for the effects of guessing. Respondents were invited to provide answers to each question and, as is the case in a multiple-choice question, some would have produced a correct answer by guessing alone. For these reasons, it is clear that we have measured *minimal* information and have been generous in deciding whether respondents possess it. In the case of major solar purchases, for example, it is difficult for rational theory to explain how people could have made these major expenditures without even the essential cost information requested in our survey.

This analysis errs in favor of the rational theory for a second reason. Our data are cross-sectional and, from the point of view of conservation device purchases, *post hoc.* Even in the infrequent cases in which respondents with solar devices were well-informed about solar tax incentives, therefore, we cannot eliminate the possibility that they became well-informed *after* their purchases. Rational theories of energy conservation imply that individuals understand relevant incentives *prior to* purchasing conservation devices, and that this understanding plays a role in calculating cost-effectiveness. Since our survey data are cross-sectional, critics of the rational model can argue the purchase led to the understanding, rather than the other way around (Ehrlich et al. 1957). From these results, we conclude that the rational model does not provide a convincing or powerful explanation of how people decide to undertake energy conservation.

## ALTERNATIVE THEORIES ABOUT ENERGY CONSERVATION BEHAVIOR

If attitudes and policy awareness are not strongly associated with conservation behavior, the simple relationships implied by attitude and rational theories of conservation do not exist. In their absence, it becomes important to ask whether there are *any* predictors that *are* linked with conservation behaviors. At the outset, many conceptions of energy conservation understate its complexity. Treatments of this subject often speak of "increasing conservation" or encouraging conservation as if a single act is involved. This simplification seems unwarranted, and Table 3 lists the diverse forms of conservation addressed in our survey. These data indicate that conservation incidence varies markedly between homeowners and renters, and across different types of conservation.

Conservation habits such as turning off lights and closing drapes and shades are performed most often. These actions require no capital investment and small inconvenience, but must be performed on a habitual basis to produce any energy savings. Other conservation habits, such as recycling and using alternate modes of transportation, are performed less often. These actions are more inconvenient and, therefore require higher levels of commitment to

**Table 3. Self-Reported Incidence Rates For Three Different Kinds of Energy Conservation**

| | Own | Rent |
|---|---|---|
| **Conservation Habits*** | | |
| Close all window drapes and shades at night | 83.1% | 85.2% |
| Turn off lights when you leave the room | 77.4 | 78.0 |
| Conserve hot water when showering, washing the dishes or clothes | 59.8 | 50.2† |
| Recycle some products | 41.7 | 36.4 |
| Use transportation other than car | 13.1 | 23.7‡ |
| **Conservation Devices**** | | |
| Insulation | 78.4 | 37.4‡ |
| Low-watt light bulbs (60 watts or less) | 74.2 | 70.5 |
| Weatherstripping | 73.2 | 44.9‡ |
| Low-flow showerhead | 57.0 | 41.7‡ |
| Water heater insulation blanket | 38.9 | 13.8‡ |
| **Solar Devices**** | | |
| Skylights | 17.5 | 7.9‡ |
| Greenhouse window | 6.2 | 2.0† |
| Solar water heater | 4.9 | 1.6† |
| Greenhouse/solarium | 4.4 | 1.6 |
| Solar heater for pool or hot tub | 3.1 | 1.2 |
| Solar space heater | 1.5 | 0.4 |
| Other solar equipment | 4.1 | 1.6 |

* Includes those who responded "almost always."

** Includes those who installed the device and those living in a home in which the device was already installed.

† These two proportions are significantly different at the 0.05 level.

‡ Proportions significantly different at the 0.01 level.

be effective.

Although there are a few exceptions, homeowners and renters report similar levels of these habitual actions. Homeowners are more likely to report that they conserve hot water, presumably because homeowners own more clotheswashers and dishwashers, and renters more often live in master-metered dwellings. Renters are more likely to use alternate modes of transportation because they are less likely to have the money to own and operate an automobile. Indeed, the median annual family income of renters in our survey was nearly $9,000 less than that of homeowners ($10,400 vs. $19,250).

Device-oriented actions, such as installing insulation or a water heater blanket, are performed less often. These "one-shot" efficiency behaviors require some initial capital investment, but do *not* require a commitment to change one's lifestyle or acquire new habits. Also, conservation devices are more likely to produce substantial energy savings than conservation habits (Stern and Gardner 1981). Once the device is installed, energy savings accrue independently of the resident's motivational level. Homeowners are much more likely than renters to install these devices, presumably because they benefit financially from home improvements while renters do not. There is only a single conservation device on which renters and owners do not differ—low-watt bulbs, which require little cost and no permanent improvement of the dwelling.

Solar devices, such as a solarium or solar water heater, are found far less frequently in our sample. This is not surprising, since many solar devices require the outlay of considerable capital. For example, the "first costs" of a solar hot water heater are approximately $2500, although tax rebates and utility incentives could make the final cost of such a system as little as $700 during the period of our survey. Other solar devices, such as skylights, are much less costly, but are still relatively uncommon in our sample. Again, homeowners are more likely than renters to take these actions, although the differences are not large.

**Three Types of Conservation Behavior**

Rather than one coherent dimension, the evidence suggests that the general rubric of "energy conservation" contains at least three independent factors. Three different measures were constructed from the items in Table 3. Table 4 shows the item orders, marginals, and coefficients for two small Guttman Scales: the Conservation Habits Scale and the Conservation Devices Scale.

**Table 4. Conservation Scales**

**Conservation Habits Guttman Scale**

| Item Description | Percentage Agreement |
|---|---|
| Item A: Recycles Products AND Turns Off Lights | 32% |
| Item B: Conserves Hot Water | 55 |
| Item C: Closes Window Drapes and Shades | 82 |

Scale Types:
4 = Items A + B + C = 19%
3 = Items B + C = 40
2 = Item C = 32
1 = No Items = 9

Total.................=100%

Coefficient of Reproducibility = 0.88
Coefficient of Scalability = 0.63

**Conservation Devices Guttman Scale**

| Item Description | Percentage Agreement |
|---|---|
| Item A: Water Heater Insulation Blanket | 29% |
| Item B: Low-Flow Showerhead | 51 |
| Item C: Home Insulation OR Weatherstripping or both | 77 |

Scale Types:
4 = Items A + B + C = 18%
3 = Items B + C = 36
2 = Item C = 32
1 = No Items = 14

Total..............=100%

Coefficient of Reproducibility = 0.88
Coefficient of Scalability = 0.66

For each measure, one of the five items listed in Table 3 was dropped, and two others were combined to form three-item cumulative scales. For the Habits Scale, the alternative transportation item revealed only low associations with the other habit items and was omitted from the Scale. This item was also strongly related to SES, with poorer, less educated respondents reporting more uses of transportation forms other than the private car. Hence, bike and bus riders in our sample may avoid the use of cars for economic reasons apart from energy considerations. Consistent with this interpretation, observe in Table 3 that this item was the only one that yielded a significantly greater response from renters than homeowners.

For the Devices Scale, the low-watt light bulbs item also was omitted because of modest relationships with the remaining items. Indeed, Table 3 shows it to be the only one of the five Devices items that failed to uncover a significant difference between renters and owners. This result may mean the item is strongly influenced by "social desirability," and the high marginals provide support for this interpretation. In addition, we may have erred by defining low-wattage at 60 watts or less; a reduced wattage definition - say, 40 watts or less - might have proved more discriminating.

These two short Guttman scales yield similar coefficients. Both have a reproducibility of 0.88, approximating Guttman's recommended standard of 0.90. For scalability, the coefficients are 0.63 and 0.66, slightly higher than the recommended standard of 0.60. With these acceptable characteristics, the Conservation habits and Conservation Devices Scales constitute two of our three primary measures of energy conservation behavior.

Solar adoption is the third conservation measure, though its items do not allow scaling. Solar devices are still uncommon even in California's sunny Santa Cruz County. Only three members of the sample of 642 reported as many as four of the solar devices listed in Table 3—a mere 0.5%. Only 14 more (2.2%) reported three devices, 31 (4.8%) reported two, and 87 (13.6%) reported one. The vast majority (79%) reported no solar devices. Moreover, Table 3 reveals that lower-cost items (skylights and greenhouse windows) predominated, with homeowners listing many more solar devices than renters. We use the Solar Devices measure in two ways: as a continuous variable with a square-root transformation (to reduce the effect of wild scores), and as a three-way categorical variable (none = 79%, one device = 13.6%, and two or more devices = 7.4%).

**Table 5. Pearson Correlations Between Three Types of Conservation By Renters and Homeowners**

| | Conservation Habits Scale | Conservation Devices Scale |
|---|---|---|
| **Renters** (N= 254) | | |
| Conservation Devices Scale | +0.01 | --- |
| Solar Devices† | 0.00 | +0.12*(+0.13)‡ |
| **Homeowners** (N = 388) | | |
| Conservation Devices Scale | +0.14**(0.16)‡ | --- |
| Solar Devices† | -0.01 | +0.14**(0.16)‡ |

* $p < 0.05$

** $p < 0.01$

† Solar devices are measured as the square root of the number of solar devices reported by the respondent.

‡ Coefficients corrected for coarse grouping (5 categories for each variable).

These three measures of energy conservation behavior (Habits, Devices, and Solar) are logically separable; they measure contrasting *strategies* of conserving energy. In addition to this conceptual distinctiveness, empirical analysis indicates the three measures are essentially independent. Table 5 provides the Pearson correlation coefficients among the three conservation measures, and these relationships are shown separately for homeowners and renters.

For both groups, there is no relationship between the Conservation Habits Scale and the measure for Solar Devices—indicating that high levels of conservation habits are not systematically associated with solar energy. This finding shows that these two major conservation strategies are unrelated, and suggests the existence of two very different orientations: one stressing the *efficiency* of conventional energy use (i.e., changing minor energy habits), and another stressing *alternative fuels* (i.e., solar).

For both owners and renters, there is a small and statistically significant positive relationship between the Conservation Devices

Scale and the Solar Devices measure. This relationship indicates that individuals who own non-solar conservation devices are slightly more likely to own solar devices as well. This finding could mean that individuals decide to acquire different types of conservation equipment simultaneously. Alternately, this finding raises the intriguing possibility that adoption of *any* conservation device may make the adoption of additional devices more likely.

The third relationship, between the Habits and Devices Scales, differs between owners and renters. Among renters, for whom conservation devices often are installed and owned by landlords, there is no correlation between the scales. Among owners, there is a small but statistically significant positive correlation between the two strategies of energy conservation. This difference reflects not only the greater income of owners but also the fact that owners are, more than renters, free to implement energy conservation. Lacking ownership, renters have few or no incentives to install many conservation devices, and this structural condition makes a consistency between habits and devices improbable for this group.

The principal message of Table 5 is not the three significant coefficients so much as the general *absence* of covariance between the measures. From this general lack of relationships, we conclude that energy conservation is best conceptualized as involving three essentially unrelated types of behavior and two contrasting populations —renter and homeowners. This framework guides the remainder of our analysis.

## PREDICTORS OF THREE TYPES OF CONSERVATION BEHAVIOR

We conducted a series of stepwise regression analyses to discover the structural, demographic, and attitudinal predictors of the three types of conservation behavior. Based on zero-order correlation matrices, a total of 34 independent variables from different survey items were entered in the initial regressions. The Guttman scales for Conservation Habits and Conservation Devices were used as dependent variables along with an additive scale (with a square-root transformation) as the measure of Solar Devices. Since these three dependent variables were analyzed separately for owners and renters, a total of six regressions were performed. Table 6 summarizes the predictors of conservation habits, Table 7 the predictors of conservation devices, and Table 8 the predictors of solar devices in the home.

The best predictor of conservation habits is the respondent's

**Table 6. Variables Predicting Conservation Habits Scale**

| | BETA | Change in R-Square | Statistics |
|---|---|---|---|
| **Homeowners** | | | |
| Self-report of effort made to conserve energy. | 0.23[†] | 0.05 | Constant= 1.34 |
| Seeks information about energy conservation. | 0.14** | 0.02 | Multiple R= 0.34 |
| Belief that utilities should give away energy-conserving devices free. | 0.13** | 0.02 | R-square= 0.11 |
| Belief that America's situation is improving. | -0.13** | 0.02 | |
| Natural log of amount of last utility bill. | 0.10* | 0.01 | |
| **Renters** | | | |
| Self report of effort made to conserve energy | 0.32 | 0.10 | Constant= 1.77<br>Multiple R= 0.32<br>R-square= 0.10 |

* $p < 0.05$

** $p < 0.01$

† $p < 0.001$

self-reported effort to conserve energy (Table 6). This result is logical but also somewhat tautological, since the habits involved obviously require some effort. For renters, self-reported effort was the only significant predictor. This variable was an important predictor for owners as well, but the picture is more complicated. For owners, a high score on the habits scale also was associated with a reported seeking of information about energy conservation; a belief that utility companies should promote conservation by distributing energy-conserving devices (e.g., water heater insulation blankets)

**Table 7. Variables Predicting Conservation Devices Scale**

| | BETA | Change in R-Square | Statistics |
|---|---|---|---|
| **Homeowners** | | | |
| SES | $0.28^{\dagger}$ | 0.08 | Constant= 1.11 |
| Seeks information about energy-conservation. | $0.22^{\dagger}$ | 0.05 | Multiple R= 0.39 |
| Household member able to do home repairs. | $0.15^{**}$ | 0.02 | R-square= 0.15 |
| Self report of effort made to conserve energy. | $0.10^{*}$ | 0.01 | |
| **Renters** | | | |
| SES | $0.34^{\dagger}$ | 0.12 | Constant= 0.22 |
| Seeks information about energy conservation. | $0.24^{\dagger}$ | 0.06 | Multiple R= 0.48 |
| Knows about financial incentive programs for energy conservation. | $0.16^{**}$ | 0.02 | R-square= 0.23 |
| Ownership of home technologies. | $0.16^{**}$ | 0.02 | |
| Self report of effort made to conserve energy. | $0.13^{*}$ | 0.02 | |

* $p < 0.05$

** $p < 0.01$

† $p < 0.001$

free of charge; and a belief that America's energy situation is worsening. Finally, the amount of the last energy bill (using a natural log transformation) is positively correlated with conservation habits.

Table 7 reveals a different pattern of predictors. While reported conservation habits are associated with various beliefs and preferences, the major predictors of device installations are structural and demographic. When owners and renters are aggregated, home ownership accounts for 14% of the variance, and socio-economic status (SES) accounts for an additional 8%—specifically, high SES homeowners are more likely to install energy-conserving devices, because these devices require a financial investment that is both more affordable and more practical for this group than for any other.

The predictors of device installations are slightly different for owners and renters. The best predictor for both groups is SES, followed by a tendency to seek information about energy conservation. High information-seekers were likely to know about the home energy audit program, to report that they read informational bill inserts, and to request a copy of our survey results. It appears that these people are interested in and favorably predisposed toward energy conservation. The presence of a household member capable of performing automotive and home appliance repairs was also predictive among owners. The availability of a "handyperson" reduces installation and maintenance costs of conservation devices and renders such devices more comprehensible. Self-reported effort made to conserve energy is the final predictor of device installations—people who have installed energy-conserving devices perceive themselves as having made a relatively large effort to conserve.

The pattern is similar among renters in that SES and information-seeking emerge as major variables. However, two additional predictors emerge for renters: knowledge of financial incentive programs for conservation and ownership of home technologies. This second variable was constructed from a question that asked about five consumer products: microwave ovens, home computers, home video games, video recorders, and hot tubs. This "hi-tech" variable can be interpreted in two ways. It could be that people who own these types of non-energy technology are favorably disposed to all forms of technical innovation, including energy conservation devices. In addition, this variable may also be a measure of a unique and highly specialized form of disposable income. The "high-tech" variable may reflect a tendency and the financial capacity for non-essential expenditures on innovative technical systems and equipment.

Table 8 summarizes the best predictors for solar devices. Consistent with network theory (Darley and Beniger 1981), the strongest predictor was whether the respondent mentioned a friend or

**Table 8. Variables Predicting Solar Devices in the Home***

| | BETA | Change in R-Square | Statistics |
|---|---|---|---|
| **Homeowners** | | | |
| Mentions friend or other personal contact as source of knowledge about solar. | 0.28‡ | 0.08 | Constant= 0.83 |
| SES | 0.19 | 0.03 | Multiple R= 0.42 |
| Ownership of home technologies | 0.16‡ | 0.02 | R-square= 0.17 |
| Accurately reports tax credits for solar. | 0.14‡ | 0.02 | |
| Number of economic disadvantages cited for use of solar. | -0.11** | 0.01 | |
| Number of uses cited for solar energy in the home. | 0.11 | 0.01 | |
| **Renters** | | | |
| Ownership of home technologies. | 0.17‡ | 0.03 | Constant= 1.01 Multiple R= 0.23 |
| Number of uses cited for solar energy in the home | 0.16‡ | 0.02 | R-square= 0.05 |

* The criterion variable for this set of analyses was the square root of an additive measure of solar devices in the home.

** $p < 0.05$

† $p < 0.01$

‡ $p < 0.001$

other personal contact as a source of information about solar. It is interesting that this "personal contact" variable is a significant predictor, while the mention of mass media information about solar was not. This finding emphasizes the pivotal importance of *social networks*, and suggests that solar technology may show a pattern of *social diffusion* similar to that observed for other innovations (Rogers and Shoemaker 1971).

The installation of many types of solar equipment requires substantial initial investment and therefore favors people with high incomes. As a result, our analysis shows that SES was the second strongest predictor of solar equipment. Ownership of home technologies, a measure we believe reflects a special form of disposable income, was the third best predictor. Again, this "high-tech" measure of disposable income suggests that people who find technology appealing in general are especially likely to be early adopters of solar technology.

The last three predictors of solar devices for owners are knowledge and opinion variables. Solar users tended to know the tax credits that are available for the installation of solar equipment. Of course, we do not know whether the solar users in our sample learned about the tax credits before or after they installed solar equipment. In addition, solar users mention few economic disadvantages of solar energy and are able to list a relatively large number of uses for solar in the home.

## BETTER MODELS FOR ENERGY POLICIES

The evidence examined in this paper suggests that two influential paradigms for changing behavior are seriously flawed and, as a result, that energy policies based on these paradigms are ill-conceived. The first of these theoretical paradigms, the attitude model, errs in its key assumption that attitudes bear fruit in behavior, and that making attitudes toward energy conservation more favorable will make conservation behaviors more likely. The second paradigm, the rational model of individual decision-making, overstates greatly the degree to which people possess and understand even the most elementary forms of vital cost information. These flaws appear to be fundamental, suggesting that policies predicated only on changing attitudes or creating incentives are likely to fail.

In addition to evidence contrary to the attitude and rational paradigms, this paper explores briefly the nature of more promising alternatives. The survey results discussed show that there are three distinct types of conservation rather than one alone. These three types of conservation behavior—habits, devices, and solar—

constitute contrasting strategies of energy conservation and are essentially unrelated. Our findings indicate that some significant correlates of conservation, such as social class and home ownership, are relatively fixed structural variables—suggesting that more than a single energy policy may be required. As an example, the factors that prompt conservation by homeowners are clearly unlike those that affect renters, and equitable energy policies must be designed to affect both groups.

The analysis also identifies other important predictors of conservation behavior, including several variables that are more promising in their implications for policy. Three examples are the findings that "high" conservers have: (1) a household member capable of making minor repairs, (2) friends or acquaintances who introduced them to energy information and devices, (3) other relatively costly technical devices and equipment. While the details of conservation programs necessarily vary across jurisdictions, these findings suggest promising policies could be developed that (1) offer to provide or arrange conservation device installation in individual homes, (2) use "social diffusion" and networks to introduce homeowners to specific conservation devices and alternative fuel devices such as solar systems, and (3) identify potential "high" conservers from the ranks of individuals who purchase other items of high technology equipment.

These are just three examples, but policies such as these would correspond more faithfully to the ways that individuals make energy decisions. As a result, they would be more likely to bear fruit in energy efficiency.

## References

Ajzen, I. and M. Fishbein. *Psychological Bulletin.* 84:888, 1977.

Anderson, R.W. and M.W. Lipsey. "Energy Conservation and Attitudes Toward Technology." *Public Opinion Quarterly* 42:17-30, 1978.

Archer, D., E. Aronson, T. Pettigrew, L. Condelli, B. Curbow, B. McLeod, and L.T. White. "An Evaluation of the Energy Conservation Research of California's Major Energy Utility Companies, 1977-1980." Report to the California Public Utilities Commission, Energy Conservation Research Group, Stevenson College, University of California, Santa Cruz, February 1983.

Condelli, L., D. Archer, E. Aronson, B. Curbow, B. McLeod, T.F. Pettigrew, L.T. White, and S. Yates. "Improving Utility Conservation Programs: Outcomes, Interventions, and Evaluations." *Energy* 184 (in press).

Crowne, D.P. and D. Marlowe. *The Approval Motive.* New York: Wiley, 1964.

Darley, J.M. and J.R. Beniger. "Diffusion of Energy-Conserving Innovations." *Journal of Social Issues* 37:150-171, 1981.

Ehrlich, D., L. Guttmann, P. Schonbach, and J. Mills. "Post-Decision Exposure to Relevant Information." *Journal of Abnormal and Social Psychology* 54:98-102, 1957.

Groves, R.M. and R.L. Kahn. *Surveys by Telephone.* New York: Academic Press, 1979.

McGuire, W.T. "The Nature of Attitudes and Attitude Change." In *The Handbook of Social Psychology*, ed. G. Lindzey and E. Aronson. Reading, MA: Addison-Wesley, 1968.

Morell, D. "Energy Conservation and Public Policy: If It's Such A Good Idea, Why Don't We Do More of It?" *Journal of Social Issues* 37:8-30, 1981.

Olsen, M.E. "Consumers' Attitudes Toward Energy Conservation." *Journal of Social Issues* 37:108-131, 1981.

Rogers, E. and F. Shoemaker. *Communication of Innovations: A Cross-cultural Approach.* New York: Free Press, 1971.

Rosenthal, R. and R.L. Rosnow, eds. *Artifact in Behavioral Research.* New York: Academic Press, 1969.

Seligman, C. and R.B. Hutton. "Evaluating Energy Conservation Programs." *Journal of Social Issues* 37:51-72, 1981.

Seligman, C., L.J. Becker, and J.M. Darley. "Behavioral Approaches to Residential Energy Conservation." *Energy and Buildings* 1:325-337, 1978.

Socolow, R., ed. *Saving Energy in the Home: Princeton's Experiments at Twin Rivers.* Cambridge, MA: Ballinger, 1978.

Stern, P.C. and G.T. Gardner. "Psychological Research and Energy Policy." *American Psychologist* 36:329-342, 1981.

Stobaugh, R. and Yergin, D., eds. *Energy Future.* New York: Random House, 1979.

Waksberg, J. "Sampling Methods for Random Digit Dialing." *Journal of the American Statistical Association* 73:40-46, 1978.

White, L.T., D. Archer, E. Aronson, L. Condelli, B. Curbow, B. McLeod, T.F. Pettigrew, and S. Yates. "Energy Conservation Research of California's Utilities: A Meta-Evaluation." *Evaluation Review* 8:167-186, 1984.

# Superinsulated Houses: The Importance of Resale Value

*Dick Holt*
*U.S. Department of Energy*

## INTRODUCTION

Extensive analyses, backed up by accurate field measurements, have shown persuasively that superinsulated homes provide a "least-cost" solution to indoor thermal comfort (Schick 1976; Shurcliff 1980; Rosenfeld et al. 1980; Dallaire 1980). Additional mortgage costs for insulation, glazing, and ventilation control may be more than offset by reduced fuel bills. Further, the investment receives favorable tax treatment through the deductibility of mortgage interest, and superinsulation techniques are easily adaptable to a wide range of climates and architectural styles. Why then is superinsulation not making more rapid inroads in the market for new houses?

A common answer to this question is that home buyers do not have good information about the economics of space-heating costs (Stern 1984). This answer may be the basis for proposed actions such as consumer information programs, demonstration houses, or house labeling. It is also possible that home buyers fear indoor air pollution, a problem that has received considerable press attention.

However, it is also possible to explain low market penetration of superinsulation techniques without appeal to ignorance or fear. It is possible to explain low market penetration by assuming that consumer reluctance to buy superinsulated houses is based on sound economic reasoning, and that they are acting rationally in their

*Energy Efficiency: Perspectives on Individual Behavior*

own best self interest.

How can this be? The explanation rests in two connected adverse factors. First, U.S. households are mobile. According to the Census Bureau, (U.S. Census Bureau 1980), the median time between moves for homeowning households in the U.S. is about seven years. Second, home buyers may have little or no confidence that they can recoup their additional superinsulation investment upon resale.

## ANALYSIS

Consider the home buyer who anticipates living in a newly purchased superinsulated home for only two or three years. An extra investment of $5,000-$10,000 will probably not have been fully repaid through fuel savings by then. Neither will much equity have been gained, since a large fraction of the mortgage payments goes to interest in the early years of a loan. The homeowner must rely almost totally on increased resale value to recoup his or her investment.

In theory, the homebuyer might expect that the investment in thermal improvement might be recovered upon resale. A regression analysis of resale data from 1317 single family houses sold in Knoxville, Tennessee in 1978 estimated that, under certain qualifying assumptions, "an investment in an energy-saving durable good resulting in a one-dollar reduction in the annual fuel bill of the house will, *ceteris paribus,* increase the market value of the house by $20.73 in 1978 dollars" (Johnson and Kaserman 1983). However, the economic model used to derive this estimate assumed that the individual discount rate applied to the energy savings over the period of occupancy is equal to the social discount rate—an assumption that may not hold true for that fraction of homebuyers who move in only a few years. The model also concedes that small values of the period of home occupancy and large values of expected lifetime of the durable good induce a "distorting effect...that may well predominate."

However, in practice, data about the resale value of thermal efficiency improvements are meager, and there is wide regional variation. The newsletter of the Society of Real Estate Appraisers reported (in 1979) that the fraction of the original cost of added insulation recovered upon resale ranged from 15%—25% in St. Louis, 20%—30% in New York City, 25%—35% in Houston, and 50%—100% in Seattle. The Remodeling Industry Trade Association (with a vested interest in estimating on the high side) characterizes added insulation as a "good investment if you plan on staying 4 or 5 years." Short term recovery estimated to be 50%." These data cannot be too reliable, since the resale values of thermal efficiency measures were not controlled for inflation in housing prices or for the value of other important house features—location, kitchen design, lot size, schools, fireplaces, etc.

Homebuyers who anticipate remaining in a house only a few years may adopt what is a perfectly rational least-cost strategy for them—pay the higher fuel bills for these few years, and avoid the risk of being stuck with an investment that may not be adequately recovered at resale. Homebuyers who plan to remain for more than two or three years may still be uncertain about when they might want to move or have to move. They also may avoid the investment, just to keep their options open.

It is thus probable that only buyers who are reasonably certain that they will live in the house long enough to recover their investment primarily through fuel savings are likely to purchase thermally efficient homes. Although such homes are economically efficient for the economy as a whole, they may not be economically efficient for the individual home buyer.

It also follows that buyers who are not economically motivated to buy thermally efficient homes will, in turn, not generate significant demand among builders, who are already just as happy to keep first costs low.

## NUMERICAL EXAMPLES

For numerical examples we draw upon data from a newly constructed superinsulated house in Vermont, analyzed extensively by Brookhaven National Laboratory (BNL) (Hagen and Jones 1983). BNL evaluated several incremental levels of investment and the resulting reductions in space heating requirements, shown below in Table 1:

**Table 1. Investments in Thermal Efficiency**

| House | Incremental Cost | Annual Heating Requirement (million Btu) |
|---|---|---|
| conventional | 0 | 77.0 |
| 5.5" walls | $2100 | 40.7 |
| 12.5" walls | $7290 | 11.7 |

We further assume a 14% mortgage rate, a 25% combined federal and state marginal income tax rate, and a national average electricity price of 7.25 cents per kWh (DOE 1983).

By several criteria, an increased investment in improved thermal efficiency should be attractive to homebuyers. For example, the $2100 investment above:

- improves the homebuyer's cash flow by reducing annual payments (principal, 0.75 x interest, plus electricity) by about $480 (more when interest deductibility is taken into account);
- provides a substantial hedge against future fuel or electricity price increases;
- has a simple payback of less that 5 years, less than the median time between moves among U.S. homeowners;
- might increase the resale value of the house substantially—by almost $16,000 according to the estimates of Johnson and Kaserman;
- receives favorable tax treatment through the deductibility of mortgage interest (even more favorable than the conservation tax credit).

And yet, homebuyers may continue to avoid such investments if they fear that it will jeopardize their net worth through failure to recapture their initial investment at time of resale.

The top half of Figure 1 shows the long-run cumulative, total out-of-pocket cost (in constant dollars) of space heat incurred by the homeowner in the absence of resale recovery of the initial investment. Over a 50-year span the $2100 investment (labeled "SI-1") results in an almost $30,000 lower out-of-pocket cost compared to the conventional (zero incremental investment) house. The $7290 investment (labeled "SI-2") results in an almost $37,000 lower total cost. In the long run, the larger investment costs less because, once the loan is repaid, fuel bills are substantially lower. Over a 50 year time span, the $7290 investment reduces space heating energy requirements from $3.8 \times 10^9$ Btu to $0.58 \times 10^9$ Btu.

In Figure 1, the vertical height of the "wedge" between respective cost lines represents the dollar value that must be recovered upon resale for the energy-efficient homeowner to break even. At some future date, the total cost lines of the energy-efficient house and conventional house cross. At that time, the owner of the

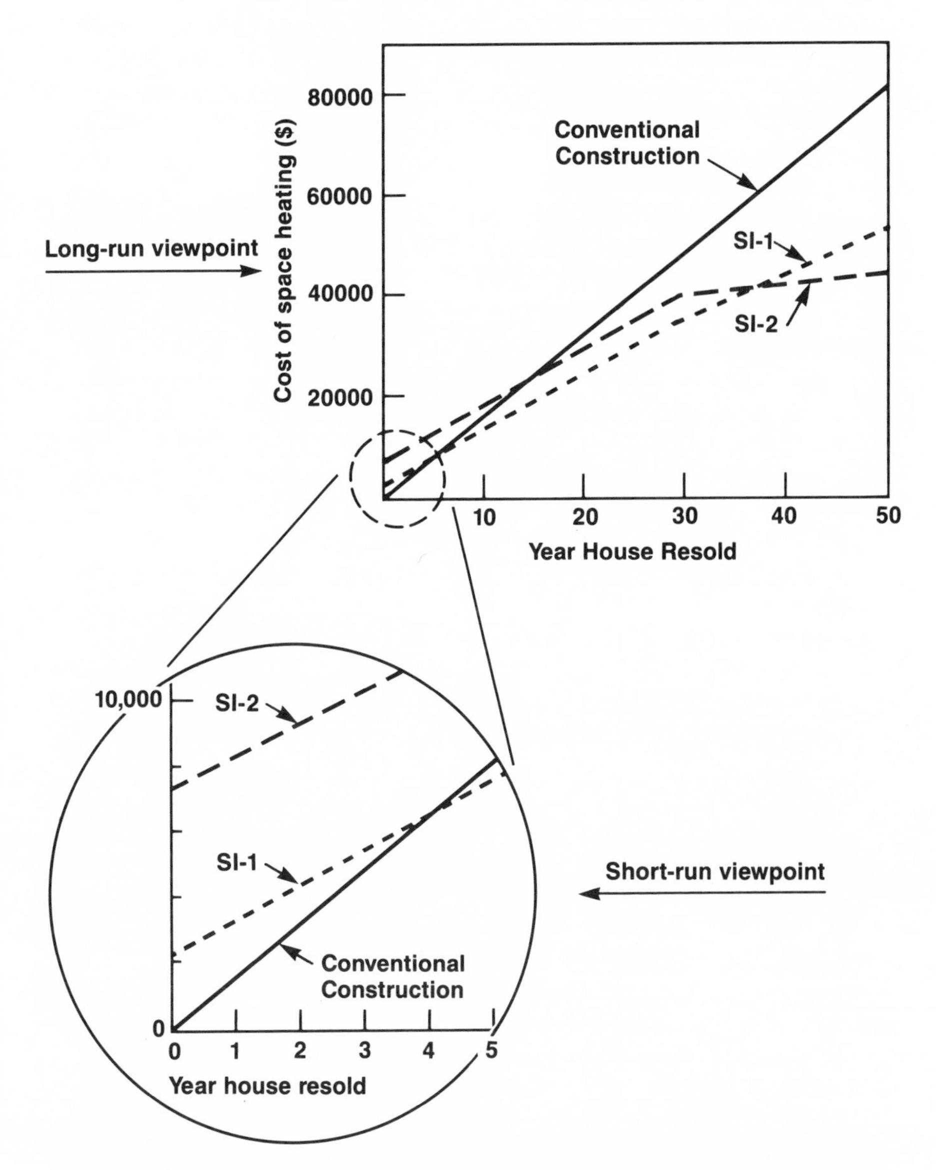


**Fig. 1. Cumulative total cost of space heating in the absence of resale recovery. The SI-1 house has 5.5″ walls and an incremental cost of $2100 compared to "conventional construction." The SI-2 house has 12.5″ walls and an incremental cost of $7290 compared to "conventional construction."**

energy-efficient house begins to turn a profit, even without any resale recovery of the initial investment. If resale value is positive, it represents a very favorable return on investment to the homeowner.

The lower half of Figure 1 contrasts the long-term and short-term views. At the top half of the figure, the 50 year time horizon shows enormous savings in money and energy for investment in superinsulation features. But most buyers will tend to look microscopically at the short-term consequences of that investment. The magnified lower half of the figure shows that, without substantial resale recovery, the short-term view will heavily favor conventional construction.

Other assumptions on fuel price, efficiency improvement costs, interest rates, etc. would, of course, produce numerically different illustrations, but the logic remains unchanged. If one were to use a present-value calculation, the lines of Figure 1 would bend downward progressively in later years, but the structure of the long-term/short-term argument would remain unchanged.

For retrofit of an existing home, I hypothesize that the effect of uncertain resale value is more severe—financing terms are likely to be more expensive, over a shorter loan period, and the construction work itself is intrinsically more expensive for a given level of improvement in the thermal integrity of the house.

## CONCLUSION

An investment in superinsulation may be economically efficient ("least cost") as seen from the longer time horizon of the nation or of the utility company. But it may not be economically efficient, in the sense of least total cost during occupancy, as seen from the shorter time horizon of the individual home buyer. From the longer range national point of view, a number of advantages accompany increased investment in thermally efficient buildings—advantages that are in part public goods:

- private savings are increased, thus tending to reduce interest rates;
- national energy consumption is reduced, thus lessening oil import dependence;
- environmental quality is improved in proportion to fuel consumption displaced;
- pressure to construct new electric generating capacity is

eased, and seasonal peaking problems are moderated.

Of these advantages, many individual home buyers may "see" only the first—increased private savings—and that only if the resale recovery of the initial investment is high in the early years. Thus, no single buyer in the sequence of buyers is likely to be willing to risk a capital loss in an investment that may primarily benefit subsequent owners. It is a challenge to those interested in the long range national welfare to devise public policies that will facilitate such private investment in thermally (and economically) efficient buildings.

**References**

Dallaire, G. "Zero Energy House: Bold, Low-Cost Breakthrough that May Revolutionize Housing." *Civil Engineering/ASCE*, May 1980.

Hagan, D.A. and R.F. Jones. "Case Study of the Blouin Superinsulated House." Brookhaven National Laboratory, BNL 51732, September 1983.

Johnson, R.C. and D.L. Kaserman. "Housing Market Capitalization of Energy-Saving Durable Good Investments." *Economic Inquiry* 21, July 1983.

Rosenfeld, A.H. et al. "Building Energy Use Compilation and Analysis (BECA): An International Comparison and Critical Review—Part A: New Residential Buildings." Lawrence Berkeley Laboratory, LBL-8912, November 1980.

Shick, W.L. et al. "Circular 2.3—The Illinois Lo-Cal House." Small Homes Council—Building Research Council, Champaign, IL, March 1976.

Shurcliff, W.A. *Double Envelope Houses and Superinsulated Houses.* Andover, MA: Brickhouse Publications, 1980.

Stern, P.C. and E. Aronson, eds. *Energy Use: The Human Dimension.* National Research Council, 1984. New York: W.H. Freeman, 1984.

U.S. Census Bureau, private communication. See also "Annual Housing Survey Part D: Housing Characteristics of Recent Movers," 1980.

U.S. Department of Energy. "Energy Projections to the Year 2010." DOE/PE-0029/2, October 1983.

U.S. Department of Energy. "Monthly Energy Review." DOE/EIA-0035 (83/12), December 1983.

# Optimizing Investment Levels for Energy Conservation: Individual Versus Social Perspective, and the Role of Uncertainty

*Ari Rabl*
*Princeton University*

## INTRODUCTION

Most energy conservation measures yield a return on an initial investment by reducing energy bills. There are many possible conservation measures, e.g. clock thermostats, attic insulation and weatherstripping, and the consumer has to decide how much to spend and which measures to install. Generally, the more one spends on conservation initially, the more one can save in the long run. But there is a point of diminishing returns, and hence one needs to determine the economically optimal investment level. For the rational consumer, that level corresponds to minimizing the life-cycle cost, i.e. the total present value of initial investment and all energy expenditures. (the relationship between future cash flows and present value involves the discount rate, a subject that is discussed in standard economics texts, for instance Riggs [1982]).

A consumer with perfect information about the actual savings due to an investment in energy conservation could optimize the investment level with certainty. But in the real world neither the future energy prices nor the magnitude of the actual energy savings

*Energy Efficiency: Perspectives on Individual Behavior*

are known in advance. Hence, the investment level is based on a "best guess" for these quantities. It is wise to consider not just the single most probable scenario but also possible variations, in order to estimate the cost penalties due to misinvestment.

This chapter provides a simple framework for evaluating these penalties. It examines the effect of misinvestment both at the level of the individual and at the level of society. For the individual, the misinvestment is due to uncertainty about the magnitude of the savings and about the future price of energy. For society there is, in addition, the mismatch between individual and social perspectives: the individual does not have to pay the full social cost of energy and his discount rate is likely to be higher than that of society (see, for example, Kempton and Montgomery (1982) on uncertainty about savings, and Hausman (1980) on individual discount rates).

A convenient measure of the cost penalty is provided by the ratio of the life-cycle costs that a consumer actually pays and the life-cycle costs that she would have paid had she chosen the optimal investment level. To evaluate this ratio one needs an explicit model for the relationship between investment level and energy savings. In this paper we consider two such models: (1) the heat flow through a layer of insulation, and (2) an exponential curve fit to a large number of residential energy conservation cost data. Both models are intuitively reasonable, and the results are qualitatively similar.

In both models the life-cycle-cost ratio can be expressed as a function of the ratio of actual energy expenditures to guessed energy expenditures. This function is simple and transparent because the individual variables (energy savings, energy prices, energy escalation rates, discount rate, and lifetime of the investment) need not be specified separately but enter only through a single quantity, the ratio of levelized energy expenditures. A plot of life-cycle cost ratio versus ratio of energy expenditures is instructive. It shows the effect not only of price uncertainties but also of uncertainties in the performance of conservation measures. The plot is equally applicable to the misoptimization problem of the individual and of society.

## LIFE-CYCLE COST OF CONSERVATION INVESTMENT

The life-cycle cost of an energy investment contains two terms, the capital cost and the energy cost. In general, higher capital costs lead to reduced energy consumption. To optimize the investment level one needs to calculate the life-cycle cost. For simplicity let us consider investments that involve only a single energy form. Let

$C$ = capital cost of energy investment
$E = E(C)$ = annual energy consumption, as function of C,

and

$p$ = price of energy.

If E and p are constant, then the life-cycle expenditure for energy is obtained by dividing the annual energy cost, (E)(p), by the capital recovery factor f, which is given by

$$f = (A/P,r,N) = \frac{r}{1 - (1 + r)^{-N}} \tag{1}$$

where
$N$ = life of investment, and
$r$ = discount rate (see for example Riggs 1982),
the notation being a helpful mnemonic for *A*nnual payment/*P*resent value. The total life-cycle cost L is the sum of the terms for capital and for energy

$$L = C + E\ p/f \tag{2}$$

When E and p change with time one can still use Eq. 2 provided one interprets E and p as levelized quantities. For example, if the energy price changes at a rate $r_p$ per year, starting from a first year price of $p_1$, then the levelized energy price is

$$p = p_1 \frac{(A/P,r,N)}{(A/P,r_p',N)} \tag{3}$$

with

$$r_p' = (r - r_p)/(1 + r_p)$$

The levelizing factor of Eq. 3 converts a sequence of changing payments into a sequence of constant payments that is exactly equivalent. (A levelizing factor can also be defined if the rate of change varies with time, but then the formula is more complicated). In an analogous manner the annual energy consumption E can be levelized. That is necessary if, for instance, the performance of a conservation measure deteriorates with time. Since E and p appear only as a product, one can treat a change or uncertainty in E as an equivalent change or uncertainty in p.

Throughout this paper we shall understand E and p to be levelized. Since L depends only on the levelized quantities, so do all the results. In particular the results are independent of the detailed manner in which E and p may change with time. The general inflation rate need not be specified because these formulas yield the same result, as long as one works consistently with either real rates or market rates.

In the following sections, the energy price and the capital recovery factor will appear only in the combination p/f, and for convenience we use a single symbol P (henceforth called life-cycle energy price).

$$P = p/f \tag{4}$$

to write the life-cycle cost as

$$L = C + EP \tag{5}$$

If the functional dependence of E on C is known, then the rational consumer can optimize the investment C such that her life-cycle cost is minimized. There is a vast variety of different energy investments, and each may have a different function E(C). In order to draw general conclusions, one needs a model that is characteristic of a large variety of investments. In this paper we consider two models that seem to be typical of conservation measures. Both are reasonable; in particular they show diminishing returns as the investment level rises.

The first model ("insulation model") is based on the loss of heat as a function of insulation thickness. Using the notation

k = thermal conductivity [W/m°C],
t = thickness of insulation [m],
A = area of insulation [m],
D = annual temperature difference-duration product or degree-seconds [°C-sec],

and

$p_{insl}$ = price of insulation material [$/m$^3$],

one obtains the annual energy flow across the insulation as

$$E = A\ k\ D/t \tag{6}$$

and the capital cost of the insulation as

$$C = A\ t\ p_{insul} \tag{7}$$

We want to vary the thickness in a situation where the area is fixed. Eliminating t in favor of C one can rewrite E in the form

$$E = K/C \tag{8}$$

where

$$K = A^2 k D p_{insul} \tag{9}$$

is a constant in a given application (for simplicity we neglect the possibility that the balance point temperature of a building and hence the number of degree days could decrease as more insulation is added). With this notation the life-cycle cost becomes

$$L(C) = C + P K/C \tag{10}$$

where P is the life-cycle energy price of Eq. 4.

Taking the derivative of L with respect to capital cost C one readily finds that the optimal investment level equals the square root of (K)(P)

$$L \text{ has minimum at } C = \sqrt{KP} \tag{11}$$

The second model is an exponential curve fit to a large number of residential energy conservation cost data. As shown by Craig et al. (1980), when one interpolates over a large set of possible conservation measures, the variation of energy consumption E(C) with investment C is well represented by the function

$$E(C) = E_o \exp(-C/C_o) \tag{12}$$

where $E_o$ and $C_o$ are constants. Now the life-cycle cost as function of C is

$$L(C) = C + P E_o \exp(-C/C_o) \tag{13}$$

Setting the derivative with respect to C equal to zero one finds the optimal investment level

$$L = \text{minimum at } C = C_o \, LN \, (P E_o/C_o) \tag{14}$$

## MISOPTIMIZATION DUE TO UNCERTAINTY

The investment is optimized according to $P_{guess}$, the best guess about p/f, but the bills must be paid according to $P_{true}$. Thus in the insulation model the actual investment level is, according to Eq. 11,

$$C_{guess} = \sqrt{KP_{guess}} \tag{15}$$

while it should have been

$$C_{true} = \sqrt{KP_{true}} \tag{16}$$

The actual life-cycle cost is

$$L_{true}(C_{guess}) = C_{guess} + P_{true}\ K/C_{guess} \tag{17}$$

while it would have been

$$L_{true}(C_{true}) = C_{true} + P_{true}\ K/C_{true} \tag{18}$$

at the true optimum. The misoptimization penalty can be expressed by the ratio

$$L_{true}(C_{guess})/L_{true}(C_{true}) = \frac{\sqrt{x} + 1/\sqrt{x}}{2} \tag{19}$$

with

$$x = P_{true}/P_{guess}$$

This ratio contains in simple dimensionless form all the information we need to evaluate the sensitivity of life-cycle cost to the misoptimization caused by uncertainty. For the insulation model it depends only on the ratio of the true and the guessed life-cycle energy prices. The life-cycle cost ratio is plotted against the energy price ratio $P_{true}/P_{guess}$ in Fig. 1a. Obviously the life-cycle cost ratio is unity when there is no uncertainty, i.e. when $P_{true} = P_{guess}$; the cost ratio increases when the guessed energy price is either higher or lower than the true price. However, the variation is rather slow. For example, even if the true energy price is twice the guessed price, the life-cycle cost is increased by only 6% beyond the minimum.

In the exponential model the life-cycle cost penalty ratio turns out to be

$$\frac{L_{true}(C_{guess})}{L_{true}(C_{true})} = \frac{x + LN(u)}{1 + LN(x) + LN(u)} \tag{20}$$

with

$$x = P_{true}/P_{guess}$$

and

$$u = P_{guess}\ E_o/C_o$$

The penalty depends on the variable $P_{guess}E_o/C_o$, in addition to the energy price ratio. It is plotted in Fig. 1b versus energy price

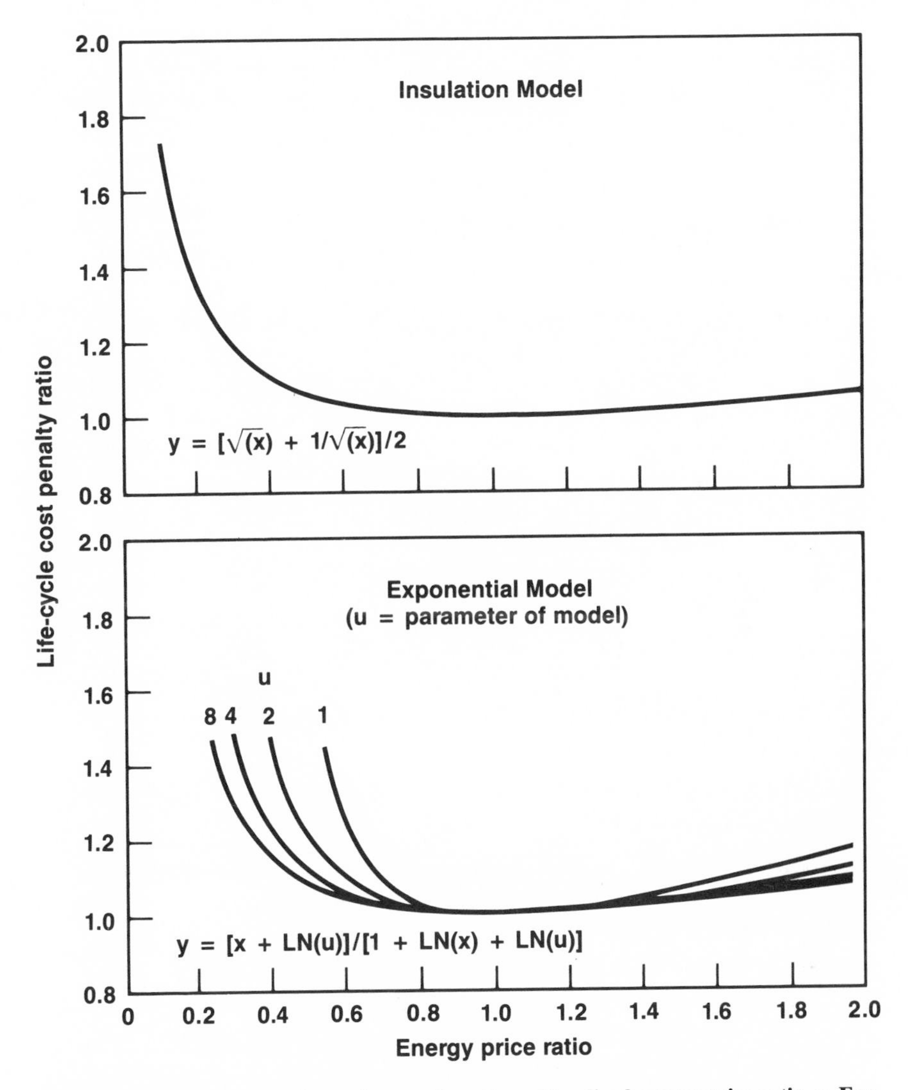


**Fig. 1. Life-cycle cost penalty ratio, y, as function of levelized energy price ratio, x. For misoptimization of individual $y = L_{true}(C_{guess})/L_{true}(C_{true})$ and $x = P_{true}/P_{guess}$. For mismatch between individual and society $y = L_{soc}(C_{ind})/L_{soc}(C_{soc})$ and $x = P_{soc}/P_{ind}$.**

ratio x, with curves corresponding to different values of the other variable. To understand this other variable, let us note that $E_o/C_o$ is the annual energy saved for the first investment dollar. The levelized annual cost of this investment is f, the capital recovery factor of Eq. 1. Hence

$$P_{saved} = f\ E_o/C_o \tag{21}$$

is the cost of saved energy for the first dollar, and

$$u = P_{guess}\ E_o/C_o = P_{guess}/P_{saved} \tag{22}$$

Certainly $P_{saved}$ must be less than $P_{guess}$, otherwise no conservation investment would be made. Also $P_{saved}$ is unlikely to be more than an order of magnitude smaller than $P_{guess}$. Hence we have plotted Eq. 20 for u = 1, 2, 4, and 8.

As with the insulation model, the curve has a broad minimum. The minimum broadens with u. Even if the true energy price is twice the guessed price ( x = 2) and for u = 1 the life-cycle cost is increased by only 18% over the minimum.

In the real world the energy/cost relationship is frequently not continuous. For instance, insulation material for walls and roofs may be available in a limited number of thicknesses. The above analysis continues to apply, but only certain points on the life-cycle cost penalty curve of Fig. 1 are physically meaningful.

## MISMATCH OF INDIVIDUAL AND SOCIAL PERSPECTIVE

Another kind of misoptimization occurs because of the difference between the perspective of the individual and that of society. The individual pays the market price of energy, not the social cost, and his discount rate is likely to be higher than that of society.

The social cost of energy is uncertain and highly controversial; however, most analysts would probably agree that in the US the present social cost is somewhat higher than the market price of energy but not more than an order of magnitude higher (Hogan 1980; Nordhaus 1980; Sweeney 1981). Instead of choosing some particular value, let us consider a range of possibilities.

Let us use subscripts "ind" and "soc" to distinguish prices and costs for the individual and for society. The individual optimizes according to his perspective, and society must pay the cost. Formally, this problem is exactly the same as the preceeding section, with the replacement

$$ind \rightarrow guess \quad \text{and} \quad soc \rightarrow true$$

Now the relevant cost penalty ratio is the social cost at the investment level of the individual divided by the social cost corresponding to the social optimum. For the insulation model it turns out to be

$$L_{soc}(C_{ind})/L_{soc}(C_{soc}) = \frac{\sqrt{x} + 1/\sqrt{x}}{2} \tag{23}$$

with

$$x = P_{soc}/P_{ind}$$

For the exponential model we obtain

$$\frac{L_{soc}(C_{ind})}{L_{soc}(C_{soc})} = \frac{x + LN(u)}{1 + LN(x) + LN(u)} \tag{24}$$

with

$$x = P_{soc}/P_{ind}$$

and

$$u = P_{ind}\, E_o/C_o$$

With this change in interpretation the curves of Figs. 1a and b are equally applicable.

## WHEN IN DOUBT, SHOULD A CONSUMER OVERCONSERVE?

Some people (Craig et al. 1980) have argued that the uncertainty of future energy prices justifies overinvesting in conservation. To shed more light on that claim, note how the optimization procedure is modified when one deals with a probability distribution of future energy prices rather than a single most probable value. First one evaluates, on the basis of one's economic outlook, the probability of occurrence of any particular value of the life-cycle energy price $P = p/f$. When this probability is combined with the associated life-cycle costs, then the optimal investment level is found by minimizing the expected value of the life-cycle cost.

The resulting investment level may be higher or lower than the level corresponding to the most probable value of $P_{guess}$. Whether it is higher or lower, i.e. whether one should overinvest or underinvest relative to $P_{guess}$, depends on the shape of the curve and on the assumed probability distribution of energy prices. For example the insulation model is symmetric under $x \rightarrow 1/x$. If the value $P_{guess} r$ is as likely as the value $P_{guess}/r$ (where r is any number) then the expected value of the penalty is minimized by investing exactly according to $P_{guess}$. On the other hand if the distribution of possible energy prices is such that $P_{guess}(1 + r)$ is as likely as $P_{guess}(1 - r)$ then the optimal investment level corresponds to a somewhat lower value of P, i.e. an underinvestment. This simple example proves that overinvestment in conservation is not always the proper response to uncertainty. Craig et al. (1980) considered only the exponential model with one particular type of energy price distribution and reached the conclusion that one should overconserve. To the extent that the exponential model is the more general of the two models, this is probably a good recommendation.

Sometimes the expected value of the life-cycle costs is not the only relevant criterion. Many investors are risk averse and place a premium on the reduction of uncertainty. An investment in energy conservation reduces the sensitivity to future energy prices. Of course, some conservation measures have their own uncertainties because performance and durability are not always known in advance. But for those technologies where the energy savings are known with sufficient certainty, overconservation will minimize the vulnerability to future energy price shocks. Therefore in practical situations overconservation may be the preferred choice.

## RELATIONSHIP BETWEEN ENERGY CONSUMPTION AND COST

It is instructive to examine how energy consumption is affected by misoptimization. In the exponential model the energy consumption is given by Eq. 12. Inserting the optimal investment level, Eq. 14 into Eq. 12, one finds that the energy consumption is

$$E = C_o/P \quad \textit{at optimum}, \tag{25}$$

the optimum being defined as the investment level that minimizes life-cycle cost when the life-cycle energy price is P. Since the individual optimizes according to $P_{ind}$, the energy consumption is $C_o/P_{ind}$, whereas it could have been $C_o/P_{soc}$ if one had optimized according to social cost. Analogous to the life-cycle cost penalty ratio we can define an energy consumption penalty ratio $E(C_{ind})/E(C_{soc})$. In the exponential model it is readily seen to be

$$\frac{E(C_{ind})}{E(C_{soc})} = \frac{P_{soc}}{P_{ind}} \tag{26}$$

Compared to the slow variation of the life-cycle cost penalty ratio, Eq. 24 and Fig. 1b, the variation of the energy consumption penalty ratio with energy price ratio is dramatic. For instance, if the social cost of energy is twice that of the individual, the energy consumption is twice the social optimum, while the life-cycle cost is only ten to twenty percent higher. A similar conclusion follows from the insulation model; here the energy consumption penalty ratio varies as the square root of the energy price ratio, but the life-cycle cost ratio also varies less than in the exponential model. Actually the fact that energy use varies strongly with life-cycle cost does not depend on a particular model but is a general consequence of the supposition that the life-cycle cost is near a minimum.

This wide swing in energy use associated with small changes in life-cycle cost has beguiled many energy analysts. It seems as if there is an enormous potential for energy savings that could be realized if only people would be willing to make more conservation investments. Such investments would increase the life-cycle cost of the individual by just a small amount, and from the perspective of society they would be justified given the current relation between social costs and market prices in the US. However, it is misleading to focus exclusively on energy use. There is nothing sacred about the level of energy consumption per se. From the point of view of society, the relevant optimization criterion is to minimize the life-cycle social cost of energy. Of course, social costs are even less well established than those for the individual. Fortunately the optimum is so broad that even a substantial error brings only a small increase in life-cycle costs.

There is another way of looking at the relationship between cost and energy: the concept of elasticity used by economists. Specifically the price elasticity of energy use $e_p$ is the relative change in energy use dE/E caused by a small relative change dp/p in energy price

$$e_p = \frac{p}{E} \frac{dE}{dp} \tag{27}$$

The elasticity has different values in the short term and in the long term. In the short term, the consumer can do little to reduce energy consumption other than to curtail her standard of living, e.g. drive less and lower her thermostat. But in the long term she can implement conservation measures such as buying a more fuel-efficient car or retrofitting her house. Therefore the elasticity of energy demand is higher in the long term than in the short term. Since we have explicit models for the costs and benefits of conservation investments, we can calculate the theoretical long-term elasticity of energy demand. Of particular interest is the elasticity at the optimum, since that is the investment level of an ideal rational consumer. Taking the derivative of the optimal energy consumption for the exponential model, Eq. 25, with respect to price and inserting it into Eq. 27, it is easy to see that

$$e_p = -1 \text{ for } \textit{exponential model.} \tag{28}$$

For the insulation model one finds

$$e_p = -0.5 \text{ for } \textit{insulation model}, \tag{29}$$

a result previously derived by Parikh and Rothkopf (1980).

These results are independent of the parameters of the models, as long as consumers are rational. The value of -1 is probably more realistic since the exponential model is a fit to a large set of conservation measures while the insulation model is only one special case. This elasticity is also consistent with the findings of Pindyck (1979) who analyzed energy consumption data and observed long term price elasticities on the order of -1. It is interesting that both models yield constant elasticity, independent of energy price. If one takes constant elasticity as a starting assumption, one can derive a more general model for the energy/cost relationship; it contains two parameters (the elasticity and an integration constant) and it interpolates between the insulation model and the exponential model.

## CONCLUSIONS

We have developed a simple and intuitive framework for evaluating the cost penalties that arise from misoptimization of energy conservation investments. The cost penalty is plotted versus the ratio of actual energy expenditures over guessed energy expenditures, the guessed value being the basis for the optimization. The graph can also be used to evaluate the cost to society due to the mismatch between individual and social perception of energy costs. Fortunately the optimum is quite broad, and the cost penalty is fairly small in most cases.

For example, if the estimate of energy expenditures is 30% high or low, the resulting cost penalty is only a few percent in most cases and even under the worst circumstances it does not exceed

10% (the latter occurs in the exponential model with u = 1 and $P_{true}/P_{guess} = 0.7$. As an illustration of the effect of uncertainties in the performance of a conservation investment, suppose the price of energy is known and the true energy savings turn out to be only 70% of the guessed savings. By the argument following Eq. 3, that case is equivalent to the situation where the energy savings are known and the true energy price turns out to be only 70% of the guessed energy price.

Our analysis also provides a new perspective on the recommendation made by Craig et al. (1980) that one should overinvest in conservation when in doubt about future energy prices. Overconservation is not always optimal, as we show with a simple counterexample, although it is likely to be a good strategy in practice.

**Acknowledgements**

*I am grateful to many colleagues who have stimulated my thinking on this subject or who have made helpful comments, especially Gautam Dutt, Eric Hirst, Willett Kempton, Robert Socolow and Robert Williams.*

**References**

Craig, P.P., M.D. Levine, and J. Mass. "Uncertainty: An Argument for More Stringent Energy Conservation." *Energy 5:* 1073, 1980.

Hausman, J.A. "Individual Discount Rates and the Purchase and Utilization of Energy-using Durables." *Bell Journal of Economics,* 33, 1980.

Hogan, W.W. "Import Management and Oil Emergencies." Kennedy School of Government report, Cambridge, MA, June 1980.

Kempton, W. and L. Montgomery, "Folk Quantification of Energy." *Energy* 7:817-827, 1982.

Nordhaus, W.D. "The Energy Crisis and Macroeconomic Policy." *The Energy Journal* 1:1, 1980.

Parikh, S.C. and M.H. Rothkopf, "Long-run Elasticity of US Energy Demand: A Process Analysis Approach." *Energy Economics,* January, p.31, 1980.

Pindyck, R.S. *The Structure of World Energy Demand.* Cambridge, MA: MIT Press, 1979.

Riggs, J.L. *Engineering Economics.* New York: McGraw-Hill, 1982.

Sweeney, J.L. EMF 6 Summary Report, Draft 3. Energy Modeling Forum, Stanford University, January 1981.

## Nomenclature

| | | |
|---|---|---|
| $C$ | = | capital cost |
| $C_o$ | = | parameter of exponential model |
| $E$ | = | annual energy consumption |
| $E_o$ | = | parameter of exponential model |
| $e_p$ | = | price elasticity of energy demand |
| $f$ | = | capital recovery factor |
| $K$ | = | parameter of simulation model |
| $L$ | = | life-cycle cost |
| $N$ | = | life of economic analysis |
| $p$ | = | energy price |
| $P$ | = | $p/f$ = life-cycle energy price |
| $r$ | = | discount rate |
| $r_p$ | = | growth rate of energy price |
| $u$ | = | $P_{guess}/P_{saved}$ or $P_{ind}/P_{saved}$ |
| $x$ | = | $P_{true}/P_{guess}$ or $P_{soc}/P_{ind}$ |
| $y$ | = | life-cycle cost penalty ratio, $L_{true}(C_{guess})/L_{true}(C_{true})$ or $L_{soc}(C_{ind})/L_{soc}(C_{soc})$ |

# SECTION III:

# Inference of Individual Behavior from Aggregate Data

# SECTION III: Inference of Individual Behavior from Aggregate Data

*Willett Kempton*
*Princeton University*

## INTRODUCTION

Although this volume concerns individual behavior, we are often obliged to use aggregate data and infer individual behavior from it. The chapters in this section use surveys, home purchase records, and other aggregate data to draw conclusions about a variety of energy-relevant individual behaviors.

Vine begins this section by demonstrating both the potential and the difficulties in using many diverse surveys to improve understanding of thermostat management. Most of the data Vine reviews were collected by utilities and state energy offices. Vine makes the case that thermostat management has significant effects on national energy consumption and that the existing data have not been fully utilized. He outlines 14 hypotheses that could guide interpretation of thermostat management data. Unfortunately, as Vine points out, the existing data are not of sufficient quality to support or reject his hypotheses. Several of the hypotheses are presented as two competing hypotheses that make opposite or conflicting predictions. These hypotheses are intended to be used not for testing, but as a basis for discussion of the existing data and as suggestions for further research. The chapter is useful both in

*Energy Efficiency: Perspectives on Individual Behavior*

suggesting ideas for further research and for illustrating the difficulty of deriving scientifically valid conclusions from heterogeneous surveys collected for diverse purposes.

In the second chapter, Laquatra uses the purchase price and the thermal integrity of new homes to infer the trade-offs made by individuals between lower operating costs and higher capital costs. As in the case of Holt's work in the first section, Laquatra asks whether the energy savings are being capitalized or, equivalently, whether future benefits (utility savings and resale price) are being converted into present value (a higher initial sale price). Since Laquatra was using data on new houses purchased in a program that might not be representative of the U.S. home buying population, he cannot directly answer Holt's question of resale. However, Laquatra does show that, in his sample, energy efficiency is capitalized at rates of return compatible with other investments available to individuals. Laquatra also makes the point that buyers need to see working examples of efficient homes to reduce uncertainty and increase the market for such homes.

Baxter and colleagues take an innovative approach to computing home energy efficiency. Using a productive efficiency model, they treat the household as a firm, and efficiency as the ratio of best observed practice to the given household's practice. Data to test the model are derived from a preexisting national survey that provides only approximations to the desired measures. Inputs are fuel costs, and outputs are number of rooms and number of occupants, controlled for climate and house type. The chapter relates differences in efficiency to physical aspects of the household and to whether the occupants rent or own their home. A comparison is also made between an efficiency analysis and a more conventional analysis of simple energy use. Contrary to what a conventional analysis might predict, households with higher efficiency are not the same as the households with lower energy use. Like Vine, Baxter et al. acknowledge that the available data make interpretation difficult. For this reason, the chapter emphasizes the method, so that it can be applied with more confidence when better data are available.

Burt and Neiman investigate the correlates of citizen support for local government actions to promote solar energy. If citizens are to support such efforts by local government, they must understand that more will be gained by pro-solar policies than will be lost by increased governmental control of the economy and of individual homes. Burt and Neiman find that citizen support of residential solar energy through requirements is related to demographic characteristics and socio-political values. Support for government promotion of solar energy is stronger among members of minority groups and the more affluent, Democratic, more environmentally conscious, better educated, and politically liberal members of the study community. Further, the authors reveal that little support exists for requiring devices such as solar water heaters and pool heaters on *existing* homes, whereas most respondents favored such requirements on *new* homes.

Finally, Morrison's chapter analyzes aggregate data of another sort. She summarizes the proceedings from an energy conservation conference that she organized: "Families and Energy: Coping with Uncertainty." Morrison focuses her synthesis on papers dealing with issues of measurement, and of monitoring techniques. She also summarizes several approaches that combine multiple methods and even finds a few papers that address theoretical issues. Morrison's conclusion suggests a need for more cross-disciplinary research combining both engineering and behavioral science approaches.

# Saving Energy The Easy Way: An Analysis of Thermostat Management[1]

*Edward Vine*
*Lawrence Berkeley Laboratory*

## INTRODUCTION

One of the most effective, least expensive, and commonly used means of reducing household energy use is to maintain low thermostat settings during the winter and high thermostat settings during the summer.[2] The monetary savings of thermostat management can be substantial: it has been estimated that $5 billion is currently saved annually in the United States due to changes in home thermostat use since the oil embargo of 1973 (Kempton 1984). Of course, this type of behavior may be merely transitory, and if people believe the energy shortage has ended, then they may keep their homes warmer in the winter and cooler in the summer, reducing or eliminating the $5 billion annual savings. This "rebound effect" may have already occurred for some households that have weatherized their homes: they may now feel that they can increase their indoor comfort level, since the owners perceive that the cost of

*Energy Efficiency: Perspectives on Individual Behavior*


[1] This chapter is a condensed version of a large report that contains a detailed analysis of the data, extensive references and tables, and an annotated bibliography (Vine 1985).

[2] For example, a 1°F increase in the summer thermostat setting can reduce cooling energy use by 4.6% in the Central Valley of California (Vine et al. 1982).

maintaining the house at a given temperature is less than before. Studying thermostat management helps to determine how much energy savings the rebound effect may have already eliminated.

Another reason for examining thermostat settings in detail is to explore the amount of variability in the way people manage their indoor comfort. Average thermostat settings are useful for modelling energy use in unoccupied or occupied homes, estimating energy use for a large sample of homes, and evaluating the impact of an energy-saving program for a utility service area, but average settings are not appropriate for estimating energy use in individual homes. Previous work in this area has shown that a few degrees difference can have a substantial effect on the energy consumed in the home. A difference of several degrees can affect the consumers' willingness to invest in energy-efficient products. Thus, knowledge of the amount of variability in thermostat settings will be useful, for example, in performing sensitivity analyses to estimate energy- and cost-effectiveness of energy retrofits for individual households.

Thermostat settings are also useful as indicators of the type of energy-saving behavior being practiced by individuals. Thermostat management is usually one of the first actions an occupant takes in reducing energy consumption in the home, and it is often the predecessor to more time consuming and expensive energy conservation measures (e.g., ceiling and wall insulation). Moreover, by examining the correlates of thermostat settings (e.g., size of a dwelling, household income, and age of the respondent), one can improve the marketing of energy-reducing programs by focusing on those variables that are highly correlated with thermostat management.

Ideally, one would like to monitor the indoor temperatures of residential households to determine if people are adjusting their thermostats to reduce energy use. However, metering thermostats is expensive and time-consuming: few studies have monitored indoor temperatures (Vine 1983). A less expensive, albeit less reliable, surrogate for measuring indoor air temperature is occupant-reported thermostat setting. The use of self-reported data raises questions about the accuracy of the data, as well as methodological issues: without objective confirmation, one does not know the veracity of an individual's reported behavior. Anecdotal data suggest a discrepancy between self-reported thermostat settings, actual thermostat settings, and indoor temperatures. So far, no one has been able to accurately estimate the relative importance of two possible sources of error—instrumentation error and behavioral error—to account for this discrepancy.[3] Also, instrumentation data are more expensive per case than survey data. Until we have a more reliable and economical method of measuring indoor temperatures, self-reported data will remain useful for improving our understanding of thermostat management.

## CONCEPTUAL MODEL AND HYPOTHESES

We analyzed data on self-reported winter and summer thermostat settings and control strategies collected in recent surveys by state and federal energy agencies, utility companies, and by our own group at Lawrence Berkeley Laboratory. We were interested not only in the distribution of thermostat settings but also in the dynamics of thermostat management. Hence, we examined how thermostat behavior was related to the following occupant-related features: socioeconomic characteristics of occupants (age, education, income, home ownership, and race), building characteristics (house type, size, and age), space conditioning fuel and system, climate, and energy audit programs. We also examined thermostat management over time (during the day, seasonally, and yearly) and analyzed its relationship to energy use. We developed a conceptual model of thermostat management to examine these variables (Figure 1). We have drawn arrows to indicate some of the possible relationships between the variables and thermostat management.

We believe that the primary sociodemographic variables affect the type, size, and age of the dwelling one occupies, which in turn affect the type of space conditioning system and fuel used in the home. The primary sociodemographic variables also affect one's probability of owning a home. The chance of receiving an energy audit is affected by many of these variables. Winter and summer thermostat settings and thermostat control are affected by all of the above variables in addition to being influenced by climate and history. Similar relationships should also affect energy use during specific periods of the day (time-of-day). Using this model, we constructed hypotheses on the relationships between thermostat management and its correlates. We developed these hypotheses on the basis of our experience with the literature on energy consumption, discussions with experts in the field, and common sense. In

[3] Instrumentation error occurs when the thermograph, which measures indoor temperature, is calibrated incorrectly. Behavioral errors often occur when respondents seek to present their most favorable image to the interviewer and provide socially desirable responses.

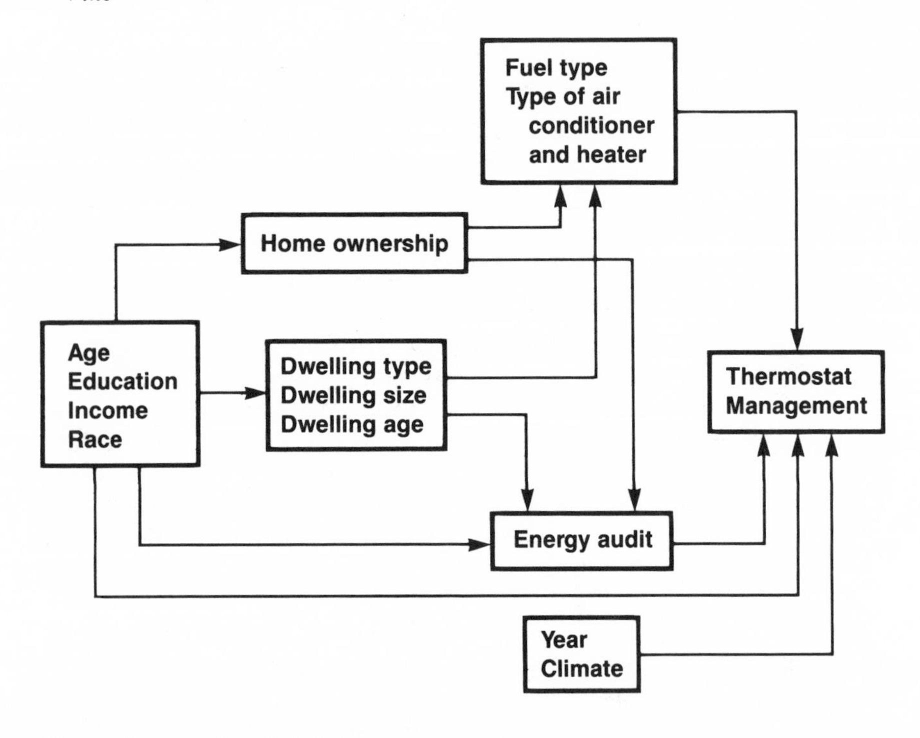


**Fig. 1. Conceptual model of thermostat management.**

several cases, we included competing hypotheses (excluding the null hypothesis) to highlight alternative relationships between variables.

1. *Age.* (a) We hypothesized that elderly people maintained higher winter settings and lower summer settings than younger people because we believed that elderly people were more sensitive than younger occupants to extreme winter and summer temperatures and were less flexible than their counterparts in adapting to a wide range of temperatures. (b) We hypothesized that elderly people maintained lower winter settings and higher summer settings than younger people because we believe that elderly people living on fixed incomes were willing to live with uncomfortable temperatures in order to reduce their utility bills.
2. *Education.* (a) We hypothesized that better educated occupants maintained lower winter settings and higher summer settings than less educated individuals because we believed that the former had more access to and knowledge of energy-saving

practices and measures. (b) We hypothesized that better educated individuals maintained higher winter settings and lower summer settings than less educated people because education was often highly correlated with income.

3. *Income.* (a) We hypothesized that higher income households maintained higher winter settings and lower summer settings than poorer households because the former could afford the cost of energy and because the latter were already using minimal amounts of heating and cooling energy and would find it difficult to cut back further. (b) We hypothesized that higher income households maintained lower winter settings and higher summer settings than poorer households because income was often highly correlated with education and with home ownership.
4. *Race.* We assumed that there were cultural norms attached to indoor comfort levels, perceived causes of illness, etc., that affected the setting of thermostats so that one might be able to distinguish racial groups, such as white and non-white households (the only ethnic group identified in the surveys). We did not know how these norms specifically affected thermostat behavior (e.g., higher or lower settings in the winter).
5. *Home ownership.* (a) We hypothesized that homeowners maintained higher winter settings and lower summer settings than renters because home ownership was often highly correlated with income and because we believed renters were more likely to adopt low-cost, energy-saving practices, such as thermostat management, than to install expensive energy-saving measures. (b) We hypothesized that homeowners maintained lower winter settings and higher summer settings than renters because the former directly received the total benefits of their energy-saving actions and were frequently the typical recipients of government and utility energy-reducing programs.
6. *Dwelling type.* (a) We hypothesized that residents of single-family houses maintained lower winter settings and higher summer settings than other residents because the former's total fuel bills were larger than their counterparts, single-family households were the typical recipients of government and utility energy-saving programs, and because we suspected fewer air leakage problems. (b) We hypothesized that residents of single-family houses maintained higher winter settings and lower summer settings than other residents because of their higher household income, and, because of their greater size, we

suspected greater air distribution problems. In addition, we believed it would be easier to maintain lower winter settings and higher summer settings for residents of apartments that capture "waste heat" from attached units.

7. *Dwelling size.* (a) We hypothesized that residents of larger homes maintained lower winter settings and higher summer settings than residents of smaller homes because of the former's higher fuel costs, ability to close off more rooms, and, because of their smaller surface area-to-volume ratio, we suspected less air leakage problems. (b) We hypothesized that residents of larger homes maintained higher winter settings and lower summer settings than residents of smaller homes because of the high correlation between size and income and the difficulty in maintaining comfortable temperatures in large homes (where air distribution posed a greater problem).
8. *Dwelling age.* (a) We hypothesized that residents of recently built homes maintained lower winter settings and higher summer settings than residents of older homes because of improved construction practices and materials (including additional insulation), and because we suspected greater air leakage and distribution problems in older homes. (b) We hypothesized that residents of recently built homes maintained higher winter settings and lower summer settings than residents of older homes because, after investing in a more energy-efficient home, the owners perceive that the cost of maintaining a new house at a given temperature is less than for an older, less efficient home.
9. *Heating fuel.* We hypothesized that electrically-heated households maintained lower winter settings than gas-heated households because of the relatively high cost of electricity.
10. *Air conditioner type.* We hypothesized that owners of room air conditioners maintained lower summer thermostat settings than owners of central air conditioners because the conditioned space was often smaller, and because we believed that owners of room air conditioners used them less than their counterparts.
11. *Energy audit.* (a) We hypothesized that audited households maintained lower winter settings and higher summer settings than before the audit and in comparison to non-audited households because the former were more knowledgeable about how to save energy. (b) We hypothesized that audited households maintained higher winter settings and lower summer settings

because, after investing in energy-saving measures, energy would be perceived to be less expensive for them than before weatherization. However, because there is a large amount of variability in the audit process–how the auditor conducted the audit, the kind of information presented, whether free or low-cost weatherization measures were installed, etc.–and because the effects of the audit may be transitory, the differences between audited and non-audited households may be negligible. Furthermore, control samples and pre-audit data are needed to accurately determine the effect of the audit on behavior.

12. *Climate.* We hypothesized that residents of cold climates maintained lower winter settings and residents of warm climates kept higher summer settings because of high fuel costs and severe climates.
13. *Year.* (a) We hypothesized that more households maintained lower winter settings and higher summer settings over time because we expected energy information and incentive programs to become more widespread and the cost of energy to increase over time. (b) We hypothesized higher winter settings and lower summer settings over time because we expected households to become more complacent and/or less interested as a result of the short-term phenomena of "energy gluts" and the rise in importance of other national issues (e.g., unemployment, inflation, and crime).
14. *Time of day.* We hypothesized that households maintained the lowest winter settings and highest summer settings at night (when they were asleep) and the highest winter settings and lowest summer settings during the evening (when people were home). During the day (when the home was often unoccupied), we expected settings to be maintained between night and evening settings.

We did not expect to confirm or disprove any of these hypotheses in this investigation. We conceived of this study as exploratory in nature, an attempt to synthesize data from diverse sources, and a quest to understand the dynamics of thermostat management.

## METHODOLOGY

Data on self-reported thermostat settings and control strategies were primarily obtained from a survey of major utility companies and all state energy conservation offices during the summer of

1983. While we recognize that this survey does not include all the utilities in the country or all research under way in academia, we feel that the survey is representative of recent thermostat management behavior in the United States. The number of households in each study numbered 50 or more. Some of the data were collected in utility customer surveys, residential energy audits, and residential energy audit evaluation surveys. In these surveys, data were collected using diverse methods: mail questionnaires, telephone interviews, and face-to-face interviews. We augmented this data base with data collected in household surveys conducted by Lawrence Berkeley Laboratory (LBL) in the past several years, in the cities of Davis and Lodi, California, and Pensacola, Florida (Cramer et al. 1985; Cramer et al. 1984; Vine et al. 1982).

Secondary data analysis is useful for evaluating the quality of primary research. Several types of problems accompany this type of analysis. Because of the diverse methods used to collect thermostat data and because of the different objectives each organization has in collecting and presenting data, it was difficult to synthesize the findings from these studies. We were dependent on what the author(s) presented, or did not present in their documents. For example, the statistical significance of the results was not reported in many of the studies that we examined, making it difficult to report definitive conclusions. Similarly, many of the reports did not contain information on missing cases for particular questions: we can only assume that most of the sample in these studies did respond to the selected questions. The importance of missing cases should not be underestimated: for example, we did not analyze data from one utility company because of the large percentage (50 to 70%) of customers not responding to several questions, although they presented a fairly thorough analysis of thermostat settings. We felt that the results presented by this utility would not have been representative of their service area.

An associated problem was the absence of thermostat data in many surveys conducted by utilities and state energy offices. Of the organizations that did collect these data, many did not present the data in their reports (i.e., the question was listed in the questionnaire without any discussion of the results in the text). And of the ones that did report the data, most of the data were presented as frequencies (without criteria of statistical significance) and rarely as cross-tabulations. Accordingly, we were left with only a few data sources for each category of thermostat settings that were of interest to us (e.g., age, income, and house size). We attempted to remedy this omission by using data from our own surveys.

## RESULTS AND DISCUSSION

We summarize our findings using the conceptual framework presented at the beginning of this chapter. Our conclusions are generally conservative and often support the null hypothesis (no relationship) when results are indeterminant.

1. *Age.* No consistent relationship seems to exist between winter thermostat settings and age, since two studies found no significant differences, one survey found lower winter settings among younger people, and a fourth study found lower winter settings among older people. Most studies found higher summer thermostat settings among younger respondents.
2. *Education.* No consistent relationship seems to exist between winter thermostat settings and education, since two studies found no significant differences, and a third survey, which found lower winter settings among less educated respondents, had serious methodological problems. Most studies found higher summer thermostat settings among higher educated respondents, although one survey found lower summer settings at night among higher educated respondents.
3. *Income.* No consistent relationship seems to exist between winter thermostat settings and income, since most studies found no significant differences, and two studies found lower winter settings among higher income respondents. Also, no consistent relationship seems to exist between summer thermostat settings and income, since most studies found no significant differences, and two studies found higher summer settings among higher income groups.
4. *Race.* The racial basis of thermostat settings and control was examined in only one report. Black households maintained warmer homes in the winter and cooler homes in the summer than white households, but black households also reduced their heating and cooling energy use by turning off their space conditioning systems.
5. *Home ownership.* No consistent relationship seems to exist between winter thermostat settings and home ownership, since one study found no significant differences, a second study found lower winter settings among homeowners, and a third study found mixed results for a number of heating practices. Home ownership was not related to summer thermostat settings.
6. *Dwelling type.* Most surveys found lower winter thermostat set-

tings among multi-family homes, although two studies found no differences. No consistent relationship seems to exist between summer thermostat settings and type of dwelling, since three studies found no significant differences, a fourth survey found higher summer settings among residents of single-family houses, and a fifth study found higher summer settings among residents of multi-family homes.

7. *Dwelling size.* The size of a dwelling was not related to winter thermostat settings or summer thermostat settings.
8. *Dwelling age.* No consistent relationship seems to exist between winter thermostat settings and age of dwelling, since most studies found no significant differences, and one survey found lower winter settings among residents of newer homes. No consistent relationship seems to exist between summer thermostat settings and age of dwelling, since two studies found no significant differences, and two studies found higher summer settings among residents of newer homes.
9. *Heating fuel.* No consistent relationship seems to exist between winter thermostat settings and heating fuel, since two surveys found no significant differences, and two studies found lower winter settings among electric-heated homes (in contrast to non-electric-heated homes).
10. *Air conditioner type.* There was only one study that examined the differences in summer thermostat settings between central and room air conditioners, and the results were inconclusive: households with room air conditioners maintained both higher and lower settings than households with central air conditioners.
11. *Energy audit.* No consistent relationship seems to exist between winter thermostat settings and energy audits, since most studies found no significant differences, although three surveys found lower winter settings among audited households. Most surveys found higher summer thermostat settings among audited households, although one study found no significant differences.
12. *Climate.* In the only study that examined the relationship between climate and thermostat settings, homes in warmer climates turned the heater off and maintained lower winter settings than homes located in other climates. The relationship between climate and summer thermostat settings was not examined in any studies.
13. *Year.* No consistent relationship seems to exist between winter

thermostat settings and year, since four studies found no significant differences. Several surveys found higher winter settings over time, and several studies found lower winter settings over time. No consistent relationship seems to exist between summer thermostat settings and year, since many surveys found higher summer settings over time, and several surveys found lower summer settings over time.

14. *Time of day.* Most surveys found significant differences in winter thermostat settings during different periods in the day, although one study found no significant differences. The typical pattern was: lowest settings at night, highest settings in the evening, and daytime settings between evening and night. No consistent relationship seems to exist between summer thermostat settings and time-of-day, since two studies found no significant differences, two surveys found lower settings as the day progressed, and three surveys found higher settings as the day progressed.

We found that thermostat behavior (especially during the summer) is clearly sensitive to certain variables (Table 1).

**Table 1. Significant Correlates of Thermostat Management**

| | Winter Thermostat Settings | | Summer Thermostat Settings | |
|---|---|---|---|---|
| Variable | Lower | Higher | Lower | Higher |
| Age | — | — | Older | Younger |
| Education | — | — | Less | More |
| Dwelling Type | Multi-family | Single-family | — | — |
| Energy Audit | — | — | Non-audited | Audited |
| Climate | Warmer | Colder | — | — |

These results strongly support three summer thermostat management hypotheses posited at the beginning of this paper (1a, 2a, and 11a) and partially support two winter thermostat management hypotheses (6b and 12). Certain groups—younger people, better educated individuals, audited households, multi-family households, and residents of warmer climates—use thermostat management to reduce energy use more often than their counterparts. Households change their thermostats during the day and during different seasons and also shut off their heating and air conditioning systems when their home is unoccupied, so the fixed thermostat settings used in many energy models and programs do not correspond with actual behavior. In fact, many households reported settings below $68^{o}$ in the winter and above $78^{o}$ in the summer.

We were unable to find consistent relationships between self-reported thermostat settings and several other variables (e.g., income, home ownership, dwelling size, and race). We experienced difficulty in interpreting the relationships between thermostat behavior and its correlates for a number of methodological reasons:

1. For a number of cases—black households, younger households, and low income groups—the sample sizes were small, making it difficult to obtain statistically significant conclusions.
2. The results were based on a single respondent's response (e.g., age, education, and thermostat setting) and may not accurately reflect how a household is actually behaving: that is, thermostat management may be a family or household decision rather than an individual decision.
3. Self-reported data are known for their methodological limitations; without objective confirmation, one does not know the veracity of an individual's reported behavior. These methodological problems may make this kind of data unreliable for statistical analysis. In fact, the self-reported incidence of energy-saving actions was reported in one study as "uniformly (and suspiciously) high", indicating a possible upward bias.
4. Diverse methods were used to collect the thermostat data (mail questionnaires, telephone interviews, and face-to-face interviews), making it difficult to synthesize the findings from these studies. Different types of samples and different sampling periods also make it difficult to arrive at a consensus.

In addition to these methodological problems, we encountered an interpretation problem: the conclusions of several studies contradicted each other, making it difficult to draw general conclusions. For example, we found higher summer settings among residents of single-family houses in one study, and, in another study, we found high summer settings among residents of multi-family homes. This indeterminacy may reflect regional differences, or it may result from competing hypotheses. Finally, for many of the hypotheses, the number of studies examined was small, leading to greater uncertainty.

We believe we need a more reliable method of measuring indoor temperatures. Advances in metering technology and computerized data collection and analysis offer the potential of measuring occupant behavior less expensively and efficiently. The problems of intervention in the household remain, and the costs are still far above survey costs, but the potential rewards are great. Metered temperature and thermostat setting data should provide a more reliable and accurate measure of indoor temperatures and thermostat management than self-reported data.

It is also important to note that one of the key differences between energy-saving practices (e.g., lowering thermostat settings) and measures (e.g., installing attic insulation) is that the former are relatively more transitory while the latter are relatively more permanent. Several studies have reported an attrition in energy-saving behavior over the last three to five years (Marylander Marketing Research, Inc. 1983; Oregon Department of Energy 1983; Tennessee Valley Authority 1983). Also, one study found that all energy-saving changes in behavior had become less common over a four year period, while all of the more permanent energy conservation investments had become more common. Nevertheless, there seems to be room for improvement in reducing energy use through air-conditioning and heating practices. Until new metering technology is more widely available, we should continue to monitor thermostat behavior to improve the accuracy of our energy models, the effectiveness of our energy programs, and our understanding of occupant behavior and energy use. To further these objectives, we plan to conduct a multivariate analysis of self-reported thermostat settings in Davis, Lodi, and Pensacola, Florida, and we intend to test the sensitivity of variations in thermostat settings on energy use using computer models developed at LBL.

**Acknowledgements**

*We would like to thank Steve Gold for his assistance in collecting the data used in this project. We would also like to thank the following for their helpful comments: Rick Diamond, Chuck Goldman, Eric Hirst, Joe Huang, Willett Kempton, Mark Levine, Jim McMahon, Max Neiman, Ron Ritschard, Mike Rothkopf, Clive Seligman, and Tony Usibelli. This work was supported by the Assistant Secretary for Conservation and Renewable Energy, Office of Building Energy Research and Development, Building Systems Division of the U.S. Department of Energy under Contract No. DE-AC03-76F00098.*

**References**

Cramer, J., N. Miller, P. Craig, B. Hackett, T. Dietz, E. Vine, M. Levine, and D. Kowalczyk. "Social and Engineering Determinants and Their Equity Implications in Residential Electricity Use." *Energy* 10(12):1283-1291, 1985.

Cramer, J., B. Hackett, P. Craig, E. Vine, M. Levine, T. Dietz, and D. Kowalczyk. "Structural-behavioral Determinants of Residential Energy Use: Summer Electricity Use in Davis." *Energy* 9(3):207-216, 1984.

Kempton, W. "Two Theories Used for Home Heat Control." In *Families and Energy: Coping with Uncertainty*, ed. B. Morrison and W. Kempton. East Lansing, MI: Institute for Family and Child Study, College of Human Ecology, Michigan State University, 1984.

Marylander Marketing Research, Inc. "1983 Conservation tracking study for San Diego Gas and Electric Company." Encino, CA, 1983.

Oregon Department of Energy. "Oregon Residential Energy Study: an Update." Salem, OR, 1983.

Tennessee Valley Authority. "1982 Interim Residential Survey: Customers of Municipal and Cooperative Distributors of TVA Power." Chattanooga, TN, 1983.

Vine, E. "Saving Energy the Easy Way: an Analysis of Thermostat Management." LBL Report 18085, Lawrence Berkeley Laboratory, Berkeley, CA, 1985.

Vine, E. "A Survey of End Use Metering in the United States." LBL Report 16322, Lawrence Berkeley Laboratory, Berkeley, CA, 1983.

Vine, E., P. Craig, J. Cramer, T. Dietz, B. Hackett, D. Kowalczyk, and M. Levine. "The Applicability of Energy Models to Occupied Houses: Summer Electric Use in Davis." *Energy* 7(11):909-925, 1982.

# Valuation of Household Investment in Energy-Efficient Design[1]

*Joseph Laquatra*
*Cornell University*

## INTRODUCTION

Although construction technology is available to greatly reduce space heating energy requirements in the residential sector, the diffusion of energy-efficient design necessary to accomplish this goal is still at an early stage of development. The fragmented nature of the housing construction industry and the uncertainties regarding energy savings and market capitalization of energy-saving durable-good investments are factors that contribute to this situation.

Homebuyer reluctance to invest large sums in energy conservation may be associated with the risk that characterizes this type of investment. The average length of occupancy for homeowners in the U.S. is 6 years (U.S. Department of Agriculture 1984), a time period much shorter than the economic lifetime of many energy-saving durable goods. A general lack of relevant information about whether residential investments in energy efficiency can be recouped through an increase in house resale value can lead households to allocate resources inefficiently in the production of ther-

*Energy Efficiency: Perspectives on Individual Behavior*


[1] This paper is based on the author's doctoral dissertation, "Housing Market Capitalization of Thermal Integrity," 1984.

mal comfort, by substituting more fuel inputs for less capital. From a policy perspective, the end result of this lack of information is the continued construction of units in the housing stock with sub-optimal levels of energy efficiency.

To address the uncertainty involved with investing in energy efficiency, government agencies have developed various incentive and demonstration programs, attempting to speed the rate of adoption of energy-conserving technologies. One such program is the Energy-Efficient Housing Demonstration Program of the State of Minnesota (Hutchinson, et al. 1982). This study will use data available from that program to address the issue of whether capitalization of energy efficiency is occurring in a specific housing market, and if so, to what extent.

The premise of this paper is that households reveal their internal rates of return for investments in energy-saving durable goods through market transactions for energy-efficient homes. From hedonic price theory (Rosen 1974), the implicit prices paid for housing characteristics can be derived from a regression of house sale price on those characteristics. Derivation of the price for energy efficiency permits computation of a rate of return on an investment in thermal integrity. Within this context, the specific objectives of this study are: (1) to determine if investments in energy-saving durable goods are capitalized into the market value of a house; and (2) to identify the internal rate of return implicit to the market value of an investment in energy efficiency under

Capitalization is the process through which future income is converted to present value. The determination of the value of specific housing characteristics, as represented by market price, is complicated by certain features of housing that separate it from other goods, including its high cost of supply, durability, heterogeneity, and fixed location (Quigley 1979). A bundle of housing attributes is indivisible, further complicating the valuation of a single characteristic, because buyers purchase an entire bundle. The marginal price of a particular item in this bundle, whether it is an additional bathroom or a qualitative aspect, is dependent on features of the bundle itself, and is therefore implicit to the bundle price. Rosen (1974) discussed the identification of this implicit or hedonic price as one that is given by the envelope of the producer's marginal reservation supply price and the buyer's marginal valuation, the bid price. In other words, at market equilibrium, the price for specific housing characteristics can be estimated from information on the entire attribute bundle.

Hedonic theory has comprised an analytical approach for studies covering a broad spectrum of issues related to the dimensionality of housing services. As described in Quigley's comprehensive review (1979), these issues have included the effect of workplace proximity on house values; the presence of externalities, such as air pollution and residential blight; taxes and public services; and estimations of income and price elasticities of demand for various housing characteristics. Other issues analyzed from this perspective were described by R. Johnson (1981), and include the value of housing quality, preference rankings of housing attributes, and racial discrimination as observed in black-white price differentials.

Recent applications of hedonic theory have examined the valuation of a structure's thermal integrity. The question of whether investments in residential energy-saving durable goods are capitalized by housing markets was investigated by Guntermann (1980), who included in his sample of 900 new homes sold in Lubbock, Texas, both superinsulated houses and houses with levels of thermal integrity as required by building codes. Using a dummy variable for energy efficiency in his regression, he concluded that energy-efficient houses sell for a 3.5% premium over conventionally constructed houses.

An examination of the housing market response to rising energy prices was conducted by Zaki and Isakson (1983). They used stepwise regression on a sample of 1,318 houses sold during a three-month period in 1978 through the Multiple Listing Service of the Board of Realtors of Spokane, Washington, and included two explanatory variables for space heating costs: (1) the price of heat, which was defined as the unit cost of heating fuel divided by the thermal efficiency of the heating equipment, and (2) a dummy variable for type of heating fuel. Neither variable was found to be significant, and the authors concluded that the relatively low price of energy in the Spokane area was responsible for these results.

In the sample of 615 gas-heated single-family dwellings sold in Columbus, Ohio, Longstreth (1981) derived capitalized values for energy efficiency using two approaches. In one model, the value of thermal efficiency was estimated from a regression that included energy-conserving features as independent variables. The second model used the quantity of natural gas consumed in a year as the variable for thermal integrity. Results from the two models led the author to conclude that some investments in energy efficiency are capitalized at higher rates than others, and that sale prices of homes in her sample were positively related to their levels of thermal integrity.

In housing markets, the existence of capitalization is evidence of homebuyer willingness to trade higher capital costs in the present for lower operating expenditures in the future. In the investment decision, the valuation of future benefits and costs is a function of the discount rate, the rate, $r$, that results in an individual being indifferent between $x$ dollars in some future time period, $i$, and $x(1+r)^{-i}$ dollars today. For example, if a person has a discount rate of 20%, she would be indifferent to receiving $100 now or $120 one year from now.

Closely related to the concept of discounting is the internal rate of return (IRR), the discount rate that equates benefits to costs. In other words, the IRR identifies, for a given investment, the discount rate beyond which net losses are incurred.[2] The IRR is a measure of profitability that depends on timing and magnitude of cash flows. As an evaluation criterion, it can be compared to the opportunity cost of capital (the expected rate of return from other investments of equivalent risk).

With an analytical framework that comprised both hedonics and capital theory, Johnson and Kaserman (1983) estimated capitalized values of energy efficiency from a sample of 1,317 houses sold in Knoxville, Tennessee. Their hedonic equation included an annual utility bill derived as an instrumental variable in a two-stage, least-squares regression. From this estimation of implicit prices, the authors reported that an investment in an energy-saving durable good that results in a utility bill decrease of $1.00 per year increases the market value of a house by $20.73.

After estimating the marginal value of energy efficiency, Johnson and Kaserman then calculated an implicit rate of return, under different assumptions regarding fuel price escalation and the remaining lifetime of the investment. They found that with a remaining lifetime for a house of 50 years and fuel-cost escalation rates ranging between 2 and 4 percent annually, the implicit housing market discount rate, or the internal rate of return, for fuel savings was between 6.3 and 8.4 percent. Comparisons between this range and bond rates on long-term U.S. government obligations in 1978 led the authors to conclude that the housing market in their

---

[2] This statement is true as long as an investment's net present value is a smoothly declining function of the discount rate. This condition holds for most residential investments in energy-saving durable goods, because benefits accumulate gradually. For more on this issue, see Brealey and Myers (1981).

study efficiently capitalized fuel savings from investments in energy-conserving durable goods.

A different approach to the question of capitalization and discounting of residential energy efficiency was taken by Corum and O'Neal (1982). They combined calculated annual heating loads of a prototype design for a house with the costs of these conservation measures under alternative applications to derive net changes in energy expenditures. Under four sets of assumptions regarding various parameters in the net-present-value calculation, rates used to discount the conservation applications were computed for ten cities. In each of the cities, savings were calculated for three fuel types: oil, electricity, and gas. A wide range of discount rates was observed among the fuel types, cities compared, and assumptions regarding financial arrangements and fuel price expectations. The authors reported a significant gap between this range and the historical range of real market interest rates. The discount rates calculated in this analysis were, in general, higher than historical market interest rates.

The studies reviewed indicate that housing-market capitalization of energy-saving durable goods is occurring, and that rates of return compare favorably with the opportunity cost of capital. To date, however, a general lack of data has prevented an examination of this issue with the use of a precise measurement of thermal integrity. With data from a state demonstration program described below, such a measurement can be used to examine the efficiency of a particular housing market with regard to capitalization of this characteristic.

## THE DATA

This study is a cross-sectional analysis of houses constructed through the Energy Efficient Housing Demonstration Program (EEHDP), which was implemented in 1980 by the Minnesota Housing Finance Agency (MHFA). Under this $11 million mortgage-loan program, 144 units were constructed with a variety of energy-efficient designs, throughout the state of Minnesota. Marginal prices for housing characteristics are not constant across markets, a factor that necessitates limiting a hedonic analysis to one area, usually defined by county lines or as a Metropolitan Statistical Area (MSA). Eighty-one of the EEHDP units are located within the Minneapolis-St. Paul MSA and comprise the sample for this study. A much larger number of observations should ideally be used for this type of research, but at this early stage of the diffusion process a large enough sample of recently sold homes with accurate measurements of thermal integrity is only available at considerable expense, and was beyond the means of this paper.

One of the program objectives of EEHDP is to gather and analyze data. For the houses built, this goal has resulted in descriptive information about their structural characteristics and thermal integrity levels. These data were supplemented with Census tract information for neighborhood variables, municipal maps, and school district data for locational attributes.

## THE HEDONIC EQUATION

In the first stage of this analysis, the marginal price for thermal integrity of a house is estimated from a hedonic equation of the general form

$$SALEPRI = f(X, DESTIF) \qquad (1)$$

where

*SALEPRI* is the sale price of the house,
*X* is a vector of structural, neighborhood, and locational characteristics, and
*DESTIF* is the design thermal integrity factor.

A low value for DESTIF (around three) indicates that a home is extremely energy efficient, while a higher value implies that a home is less efficient.

A linear specification of the hedonic equation was indicated from a test for functional form described by Zaki and Isakson (1983). In a test for heterosedasticity on the Ordinary Least Squares (OLS) model (Breusch and Pagan 1979), the null hypothesis of homosedasticity was rejected. Accordingly, a Weighted Lease Squares (WLS) regression was selected for the final estimation.

Variable definitions and results from the WLS regression are shown in Table 1. The $R^2$ for this regression indicates that over 67 percent of the variation in SALEPRI is explained by this set of variables. Root Mean Squared Error (RMSE) for the WLS regression of 206 indicates superior predictive ability over the OLS model (RMSE = 3,323). The coefficient for DESTIF, the variable of interest, indicates that a unit decrease in the thermal integrity factor results in an increase in house price by $2,510.

**Table 1. Implicit House Characteristic Prices.**

| Variable Name | Definition and Units | Parameter Estimate | Statistic |
|---|---|---|---|
| SALEPRI | House sale price ($) | — | — |
| FLRAREA | Floor area in house (sq. ft.) | 2.81 | 0.90 |
| LTSIZ | Size of lot (sq. ft.) | 0.05 | 0.54 |
| DUPLX | Dummy variable for duplex unit | 2,979 | 1.88 |
| ATCHD | Dummy variable for attached unit | -8,641 | -5.16 |
| DESTIF | Design Thermal Integrity Factor (BTU/sq.ft./DD) | -2,510 | -2.44 |
| MEDVAL | Median house value | 0.81 | 10.93 |
| PUPEXP | School district expenditures/pupil ($) | 12.90 | 8.27 |
| WORKJOUR | Mean journey-to-work time for census tract (minutes) | -944 | -4.20 |
| INTDIST | Distance from interstate ramp (miles) | 1,645 | 5.99 |
| | $R^2 = 0.6722$ | | |
| | RMSE = 206 | | |

## IMPLICIT RATES OF RETURN

Results from the WLS regression indicate that in this sample of demonstration houses, investment levels in energy efficiency that resulted in a marginal increase in thermal integrity were capitalized into house sale price. Results from this model are used to calculate the value of thermal integrity for each house in the sample, in a comparison with the highest DESTIF of the group, which is 3.0. This "high" thermal integrity factor is much better than the average for new construction in Minnesota, which is between 6.0 and 8.0 (Hutchinson, *et al.*, 1982). The calculation used will be useful for observing the effect of investing in levels of energy efficiency that are substantially beyond that which is typical for new construction. The cost of the investment to the new home purchaser in the $j^{th}$ unit ($CST_j$) is then calculated as

$$CST_j = (3.0 - DESTIF_j) \times 2{,}510 \qquad (2)$$

The net-present value of the investment in energy efficiency is equal to the net savings in heating expenditures, less initial and amortized incremental costs, plus the resale value of the investment and associated tax credits and net tax savings. In a rate of return framework, this relationship can be expressed as

$$\begin{aligned} 0 = {} & [(3.0 \times FLRAREA_j \times 8195) \times FUELPRI_O] \sum_{n=1}^{N} (1+f)/(1+r)^n \\ & - [(DESTIF_j \times FLRAREA_j \times 8195) \times FUELPRI_O] \\ & -0.25(CST_j) - [0.75(CST_j) \times i(1+i)^N/((1+i)^N - \\ & [((1+r)-1)/r(1+r)^N] + CST_{jN}/(1+r)^N - \sum_{n=1}^{N} \tau CST_j(1+\pi)^n \\ & + TC_j/(1+r) + \sum_{n=1}^{N} [T_{jn}/(1+r)^n] \end{aligned} \qquad (3)$$

where

| | | |
|---|---|---|
| 3.0 | = | Highest thermal integrity factor in sample |
| $FUELPRI_O$ | = | Price of fuel in \$/BTU at time O |
| $N$ | = | Length of mortgage (30 years) |
| 8195 | = | Minneapolis heating degree days (65°F base) |
| $f$ | = | Fuel cost escalation rate |
| $r$ | = | Internal rate of return |
| $\pi$ | = | The rate of annual property value appreciation |
| $T$ | = | Property tax rate |
| $TC_j$ | = | State tax credit: 20% of CST, limited to first \$10,000 |
| $T_{jn}$ | = | Tax savings for the $j^{th}$ household in year n, calculated as in Equation 4. |

The deductibility of mortgage interest and property tax payments favorably affects the household's valuation of an investment in real property. For the $n^{th}$ year, tax savings are

$$T_{jn} = t_{jn} [I_{jn} + \tau CST_j (1+\pi^{n})] \qquad (4)$$

where

$t$ = marginal tax rate
$I$ = interest payment, calculated from a mortgage program, using (0.75) $ST_j$ as the principal.

The heating fuel for units in this sample is natural gas. Fuel prices are expressed on a per BTU basis (prices were obtained from the Minnesota Department of Energy, Planning, and Development). Sixty-one percent efficiency was assumed for gas heating systems. Property tax rates for the school districts represented in the sample were adjusted for Minnesota's Homestead Exemption. State confidentiality regulations prohibited the release of all but summary income statistics for the households in the sample. The sample was divided into 12 groups according to house type, and the mean income for each group was calculated. The income figure used for each household is the mean for its respective group. This figure, household size, and 1980 federal income tax tables were used to derive a marginal tax rate for each household. These rates increased each year as $(1+\alpha)t_{j(n-1)}$. Based on a ten percent annual per capita income increase calculated from Minnesota historical data (Economic Development Administration 1977; Bureau of the Census 1983), $\alpha$ was set at 0.10. Three alternative scenarios projected by the U.S. Department of Energy (1983) were used to calculate fuel cost escalation rates. In these projections, the price of fuel changes every five years. The lowest escalation occurs in A, with an average increase of 0.10 every 5 years; in B the average increase is 0.15; and in C it is 0.22.

Using the results of the hedonic regression to calculate $CST_j$ as in Equation 2, an iterative computational procedure was used to solve for r in Equation 3, under the three energy-price-escalation scenarios. The use of mean incomes for groups defined according to house type resulted in rates of return that differed within groups according to the household marginal tax bracket. Accordingly, representative observations from the groups were chosen for inclusion in Table 2, where results of the iteration are summarized, and assumptions for the different cases are listed. Characteristics of the units represented are shown in Table 3.

Following M. Johnson (1981), $\pi$ was set at 0.10 in Cases I, III, and IV, as a basis for comparison against a zero rate of property value appreciation in Cases II and V. The full thirty-year mortgage period is compared with six years of ownership in Cases IV and V.

**Table 2. Implicit Internal Rates of Return.**

| UNIT | I | | | II | | | III | | | IV | V |
|---|---|---|---|---|---|---|---|---|---|---|---|
| IDENTIFIER | A | B | C | A | B | C | A | B | C | B | B |
| 1 | 0.16 | 0.26 | 0.39 | 0.16 | 0.27 | 0.39 | 0.16 | 0.26 | 0.39 | 0.09 | 0.06 |
| 2 | 0.16 | 0.26 | 0.39 | 0.16 | 0.26 | 0.39 | 0.16 | 0.26 | 0.39 | 0.09 | 0.06 |
| 3 | 0.17 | 0.27 | 0.40 | 0.17 | 0.27 | 0.40 | 0.17 | 0.27 | 0.40 | 0.10 | 0.07 |
| 4 | 0.16 | 0.26 | 0.39 | 0.16 | 0.26 | 0.39 | 0.16 | 0.26 | 0.39 | 0.09 | 0.06 |
| 5 | 0.17 | 0.27 | 0.39 | 0.17 | 0.27 | 0.40 | 0.17 | 0.27 | 0.39 | 0.10 | 0.07 |
| 6 | 0.16 | 0.26 | 0.38 | 0.16 | 0.26 | 0.38 | 0.15 | 0.25 | 0.38 | 0.08 | 0.05 |
| 7 | 0.16 | 0.26 | 0.39 | 0.16 | 0.26 | 0.39 | 0.16 | 0.26 | 0.39 | 0.09 | 0.06 |
| 8 | 0.16 | 0.26 | 0.39 | 0.17 | 0.27 | 0.39 | 0.16 | 0.26 | 0.39 | 0.09 | 0.06 |
| 9 | 0.16 | 0.26 | 0.39 | 0.16 | 0.26 | 0.39 | 0.16 | 0.26 | 0.39 | 0.09 | 0.06 |
| 10 | 0.20 | 0.31 | 0.44 | 0.20 | 0.31 | 0.44 | 0.20 | 0.31 | 0.44 | 0.11 | 0.09 |
| 11 | 0.19 | 0.29 | 0.42 | 0.19 | 0.29 | 0.42 | 0.19 | 0.29 | 0.42 | 0.11 | 0.08 |
| 12 | 0.19 | 0.29 | 0.42 | 0.19 | 0.29 | 0.42 | 0.18 | 0.29 | 0.42 | 0.10 | 0.08 |
| 13 | 0.20 | 0.30 | 0.43 | 0.20 | 0.30 | 0.43 | 0.19 | 0.30 | 0.43 | 0.11 | 0.09 |
| 14 | 0.18 | 0.29 | 0.42 | 0.18 | 0.29 | 0.42 | 0.18 | 0.29 | 0.42 | 0.10 | 0.07 |
| 15 | 0.19 | 0.30 | 0.43 | 0.19 | 0.30 | 0.43 | 0.18 | 0.30 | 0.42 | 0.11 | 0.08 |

**Legend:**

I: N = 30, $\pi$ = 0.10
II: N = 30, $\pi$ = 0
III: N = 30, $\pi$ = 0.10, Resale Value = 0
IV: N = 6, $\pi$ = 0.10
V: N = 6, $\pi$ = 0

**Table 3. Constants in the Calculation**

| UNIT IDENTIFIER | DESTIF | CST | t |
|---|---|---|---|
| 1 | 1.24 | $4,418 | 0.14 |
| 2 | 1.30 | 4,267 | 0.14 |
| 3 | 1.43 | 3,941 | 0.20 |
| 4 | 1.55 | 3,640 | 0.15 |
| 5 | 1.55 | 3,640 | 0.20 |
| 6 | 1.60 | 3,514 | 0.13 |
| 7 | 1.60 | 3,514 | 0.19 |
| 8 | 1.80 | 3,012 | 0.15 |
| 9 | 1.85 | 2,886 | 0.13 |
| 10 | 1.95 | 2,636 | 0.13 |
| 11 | 2.01 | 2,485 | 0.20 |
| 12 | 2.19 | 2,033 | 0.14 |
| 13 | 2.19 | 2,033 | 0.21 |
| 14 | 2.48 | 1,305 | 0.11 |
| 15 | 2.48 | 1,305 | 0.20 |

The effect of a zero resale value of the investment in thermal integrity is observed for the 30-year ownership period in Case III. A case not included in the table, which will be discussed, is a zero resale value in the 6-year time horizon.

The fuel price escalation rates change every five years in the Department of Energy projections, making the sixth year the first year with a price change. One year of different fuel prices did not produce savings substantial enough to affect the discount rate across the three scenarios, in the six-year time horizon. Changes may have occurred at or beyond the third decimal place, but the computational precision of the iterative program used was to two decimal places. For this reason, in the six-year ownership period, only the results from Scenario B are presented.

## DISCUSSION OF RESULTS

As seen in Cases I, II, and III, the rates of return are generally insensitive to the changes made in assumptions regarding property value appreciation ($\pi$) and resale value. Valuation changes affect both net property tax payments and resale value. With an effective assessment rate of 0.26 used in this analysis, changes in net tax payments resulting from a change in $\pi$ were negligible, although slight increases in the rates of return are seen for observations 5 and 8. A slight decrease in the rate of return, resulting from a zero resale value, is seen for observation 6, 12, and 15.

Another noticeable feature of the implicit rates of return in the thirty-year projections is that they rise with increases in fuel prices, indicating the effect of higher savings in energy expenditures. Table 3 illustrates the effect of a change in the marginal tax rate. Pairs of observations that are identical except for this rate are 4 and 5, and 7, 12 and 13, and 14 and 15. Generally, an increase in the marginal tax rate results in an increase in the rate of return. For example, in Case I under Scenario A, rates of return are 16 and 17 percent for observation 4 and 5 respectively. Respective marginal tax brackets for these observations are 15 and 20 percent. Rates of return are positively related to marginal tax brackets because of the value of income tax deductions for mortgage interest and property taxes.

Rates of return derived under the 6-year time horizon are much lower than those calculated for the thirty-year ownership period. Sensitivity to a change in $\pi$ is seen in Case V. The decrease in the rate of return results from the effect of $\pi$ on the resale value of the investment, which has more of an effect when discounted from six years than from thirty years. This effect was also apparent when the resale value was set at zero, a change that resulted in there being no solution for an internal rate of return. This last result reflected the effect of net negative cash flows.

Available measures of the opportunity cost of capital, for comparison with the results of the 30-year analysis, are the interest rate on long-term U.S. Government bonds, which was 14.52 percent in July 1981 (the midpoint of the mortgage closing dates for this sample), and the market rate of interest on conventional mortgages, which was 17.50 percent at that time (Board of Governors of the Federal Reserve System 1981). The results under Scenario A compare closely with these measures and exceed them in Scenarios B and C. For example, the range of rates is from 15 to 20 percent in Scenario A and exceeds 40 percent in Scenario C. This result indicates that the market where these transactions took place has capitalized the future value of net benefits associated with these investments in thermal integrity. The rate of return can be thought of as the discount rate applied to an investment that results in a zero net percent value, or equivalent benefits and costs. When the discount rate exceeds the interest rate on funds borrowed to finance the investment, a profit is made.

The lower rates of return derived under the six-year time horizon result from a truncation of net benefit streams. When the rate of property value appreciation is set at 0.10, all derived rates are higher than the mortgage interest rate. This is not the case when the value of this parameter is set at zero, indicating a higher sensitivity of the results to this parameter in the shorter ownership period. The fact that EEHDP occupants are first-time buyers highlights the significance of this result. As first-time buyers, they are likely to move within the six-year period, meaning that the thirty-year time frame may not be appropriate for this sample. Of course, this result implies that the below market mortgage interest rate will not affect patterns of residential mobility. Further research on these buyers would be necessary to analyze this issue.

## CONCLUSIONS AND IMPLICATIONS

One observation from this study is that, for the houses in this sample, rates of return and the design thermal integrity factor (DESTIF) are positively related to each other. In other words, the higher the thermal integrity factor (or the lower the level of energy efficiency), the higher the rate of return. An implication of this result is that incentives for investing in energy efficiency may be more effective if they are tailored to the size of the investment. Equity concerns could also justify a sliding scale based on individual marginal tax brackets, as changes in rates of return resulting from tax deductions vary directly with the income tax rate, so that the higher one's income, the higher the value of the tax deduction.

The sensitivity of the rates of return to parameters affecting resale value in the six-year ownership period highlights the need for further research on this aspect of energy-saving durables. Given that this variable has a substantial effect on the rate of return, homeowner interest in quick payback times for conservation investments is justified. In the aggregate, uncertainty about resale values of these investments results in unrealized savings in expenditures for fossil fuels.

The issue of uncertainty poses a major obstacle to routine inclusion of energy-efficient design in new construction. The process of change in the housing industry is slow. As observed by Farhar-Pilgrim and Unseld (1982), potential buyers need to see working examples of energy-efficient homes. Demonstration programs such as EEHDP are important for the documentation of technological and economic feasibility. Builders and buyers alike become educated about conservation technology, not only through direct participation in the program, but also through the diffusion process.

**References**

Board of Governors of the Federal Reserve System. *Federal Reserve Bulletin* 67(12), December 1981.

Brealey, R. and S. Myers. *Principles of Corporate Finance.* New York: McGraw Hill, 1981.

Breusch, T.S. and A.R. Pagan. "A Simple Test for Heterosedasticity and Random Coefficient Variation." *Econometrica* 47(5):1287-1294, September 1979.

Bureau of the Census. *County and City Data Book, 1983.* Washington, DC: U.S. Government Printing Office, 1983.

Corum, K.R. and D.L. O'Neal. "Investment in Energy-Efficient Houses: An Estimate of Discount Rates Implicit in New Home Construction Practices." *Energy,* 7(4):389-400, 1982.

Economic Development Administration. *Selected Statistics on the Developing USA.* Washington, DC: U.S. Department of Commerce, May 1977.

Farhar-Pilgrim, B. and G.T. Unseld. *America's Solar Potential.* New York: Praeger, 1982.

Guntermann, K.L. "Energy Efficient Housing: Costs and Market Prices." Paper presented at the Allied Social Science Association Meeting, September 5, 1980.

Hutchinson, M., M. Fagerson, and G. Nelson. "Measured Thermal Performance and the Cost of Conservation for a Group of Energy Efficient Minnesota Homes." Paper presented at the Second ACEEE Summer Study on Energy Efficiency in Buildings, Santa Cruz, CA., 1982.

Johnson, M.S. "A Cash Flow Model of Rational Housing Tenure Choice." *AREUEA Journal* 9(1):1-17, Spring 1981.

Johnson, R.C. "Housing Market Capitalization of Energy Saving Durable Good Investments." Oak Ridge National Laboratory, Oak Ridge, TN, ORNL/CON-74. July 1981.

-------- and D.L. Kaserman. "Housing Market Capitalization of Energy-Saving Durable Good Investments." *Economic Inquiry* 21, 1983.

Longstreth, M. "The Capitalization of Energy Efficiency in Housing Prices." Ph.D. dissertation, Ohio State University, 1981.

Quigley, J. M. "What Have We Learned About Urban Housing Markets?" In *Current Issues in Urban Economics*, ed. P. Mieszkowski and M. Straszheim. Baltimore: Johns Hopkins University Press, 1979.

Rosen, S. "Hedonic Prices and Implicit Markets: Production Differentiation in Pure Competition." *Journal of Political Economy* 82:34-55, 1974.

U.S. Department of Agriculture, Extension Service. *Energy and Environment.* Washington, DC: U.S. Department of Agriculture, April 1984.

U.S. Department of Commerce, Bureau of the Census. "1980 Census of Population and Housing. Census Tracts. Minneapolis-St. Paul, MINN-WIS. Standard Metropolitan Statistical Area." PHC80-2-244, August 1983.

U.S. Department of Energy, Office of Policy, Planning, and Analysis. "Energy Projection to the Year 2010: A Technical Report in Support of the National Energy Policy Plan." DOE/PE-0029/2, October 1983.

Zaki, A.S. and H.R. Isakson. "The Impact of Energy Prices: A Housing Market Analysis." *Energy Economics* 5(2):100-104, 1983.

# Determinants of Household Energy Efficiency and Demand

*Lester Baxter*
*California Energy Commission*
*Stephen Feldman, Arie Schinnar, and Robert Wirtshafter*
*University of Pennsylvania*

## INTRODUCTION

One recurrent message from other chapters in this volume is that individual energy-use patterns differ quite widely. In order to develop better policy, researchers seek explanations for these variations. The most common approach has been to construct equations that evaluate the demand for energy or expenditures on energy as a function of a series of characteristics about the home and its occupants. From this type of analysis, those variables related to greater consumption of energy, or greater expenditures on energy, are singled out as possible targets for improvement. While studies of energy demand have provided insight into the motivation of individual energy users, such measures may not be the most important focus for policy. Ultimately, the amount of energy consumed is secondary to how efficiently the energy is being used.

In this chapter, we conduct and compare two different analyses of the same energy use data. In the first analysis, we use a new measure of efficiency that we introduced elsewhere (Baxter et al.

*Energy Efficiency: Perspectives on Individual Behavior*

1986) to estimate household energy efficiency in 1979. The second analysis applies one measure of energy use, total energy expenditures, used in many conventional analyses of energy demand. We compare the results of the two analyses and follow with a discussion of the interpretive differences the two approaches provide. The implications of these differences for researchers and policymakers are also considered.

Our approach to efficiency measurement treats energy as an input used for the production of household services. Those households using the fewest inputs per unit output define an efficiency frontier to which all other households' relative efficiencies are compared. This method is an extension of early work by Farrell (1957) in the measurement of productive efficiency of firms, as well as later generalizations of the approach by Charnes, Cooper, and Rhodes (1978). Our computation is based on the algorithmic procedure developed by Schinnar (1980b).

Data from the 1979 Household Screener Survey (HSS), a sample of households from the coterminous United States, are used to construct a typology of households based on attributes such as location, type of dwelling unit, and primary house heating fuel. The resulting 207 household types represent a mainstream set of household types in the continental U.S. Efficiency scores are then computed for each household type.

A regression of the efficiency scores on a host of variables reflecting locational, dwelling unit, and occupant characteristics is performed to identify those factors that account for the differences in efficiency between types of households and to control for biases in the specification of the efficiency model. The results of our study indicate that electrically heated homes are more efficient than those heated by natural gas, that multiple-fuel-use homes with higher ratios of nonelectric to electric fuel inputs are more efficient, and that owner-occupied dwellings are more efficient than rented properties.

The second analysis is conducted to determine if the relationships noted above result either from the questions asked (that is, the selection and formulation of variables) or the methods applied to address the questions (the efficiency measure used). Total household energy expenditure is substituted for the efficiency measure as a dependent variable in the regression analysis. A comparison of the coefficient estimates for the two models reveals that homes with electric heat, higher ratios of nonelectric to electric fuel inputs, and owner-occupied dwellings are all associated with higher efficiency, but not lower energy costs. This comparison suggests that units with higher efficiencies have different characteristics than units with lower energy expenditures.

## MEASUREMENT OF HOUSEHOLD ENERGY EFFICIENCY

This section describes the Best Practice Frontier Methodology used to construct an index of household energy efficiency, and reviews the household typology developed to characterize the data used in the efficiency analysis. A production function for household services requiring energy is estimated and efficiency scores are computed for each household in the typology. The efficiency score is used as a dependent variable that is analyzed in the third section by a regression model. The results of this efficiency analysis are then compared with the estimation of a household energy demand equation.

### The Best Practice Frontier Methodology

The level of fuel use by a household is a function of the demand of its occupants for household services (such as heating and cooling of interior spaces, water heating for showers, dish washing, and clothes washing, lighting for work and recreation, and so on) and the efficiency with which various energy-using equipment converts fuel inputs into the desired services. Viewed from this perspective, the household may be treated as a producer of goods and services for its occupants.[1] The dwelling unit and its energy-using equipment are the fixed (capital) factors of production, and the fuel inputs are the variable factors. In the short run, the level of a particular household service, such as house heating, may be varied by modifying the fuel input. In the long run, the level of service may be modified by adding insulation or by the installation of a more efficient heating system, altering both the variable and fixed factors of production.

The methodological approach to measuring household energy efficiency entails two steps. First, the set of households forming a frontier of best practice in energy use is identified. Second, the distance of all other households from this frontier is measured. The data envelopment technique used to carry out these steps results in

---

[1] Scott (1980) and Quigley (1984) adopt similar frameworks in their studies of household energy use.

an index reflecting the efficiency of energy use by households. This index will then be regressed on a group of variables reflecting household demographic and economic characteristics as well as physical and locational attributes of the dwelling unit in order to account for the differences in efficiency between households.

The method used to measure efficiency is based on the Charnes, Cooper, and Rhodes (1978) data envelopment technique, which is a generalization of Farrell's (1957) measure of the technical efficiency of economic production. Schinnar's (1980b) algorithmic extension of this method is the measurement tool for the empirical analysis to follow.

The approach to measuring efficiency taken here compares a group of productive units (households) along several dimensions of input resources and service outputs. The comparison identifies a subset of households that are considered efficient because they require the fewest resources per unit of service provided. These efficient households form a production possibility frontier or best practice frontier, and the index of household efficiency is based on their distance from the frontier.

The application of regression analysis to the efficiency scores enables us to introduce other variables to the efficiency analysis that are not properly considered inputs or outputs, yet are related to energy efficiency. These include such factors as climate, fuel prices, fuel availability, type of dwelling unit, etc.

To illustrate our approach to efficiency measurement, consider a household service provision process, such as water heating, requiring two inputs and resulting in a single output. Figure 1 shows six households ($H_1$ through $H_6$) distributed in an input-per-unit-output space and the location of the best practice frontier in that space. Each axis represents the input level per unit output. For example, $X_1/Y$ is electricity use per gallon of hot water generated while $X_2/Y$ is the capital cost of the water heater per gallon of hot water generated. Households $H_1$ and $H_2$ are more efficient than $H_5$ because they use less inputs than $H_5$ per unit output. Similarly, $H_2$ and $H_6$ and are more efficient than $H_3$ because they use less input per output provided. The line segments combining $H_1$, $H_2$, and $H_6$ form a piecewise linear efficiency frontier that represents a best practice pattern of service provision. Households $H_3$, $H_4$, and $H_5$ are inefficient relative to that frontier since households $H_1$, $H_2$, and $H_6$ (or a combination thereof) have demonstrated that it is possible to produce the same level of output with fewer resources.

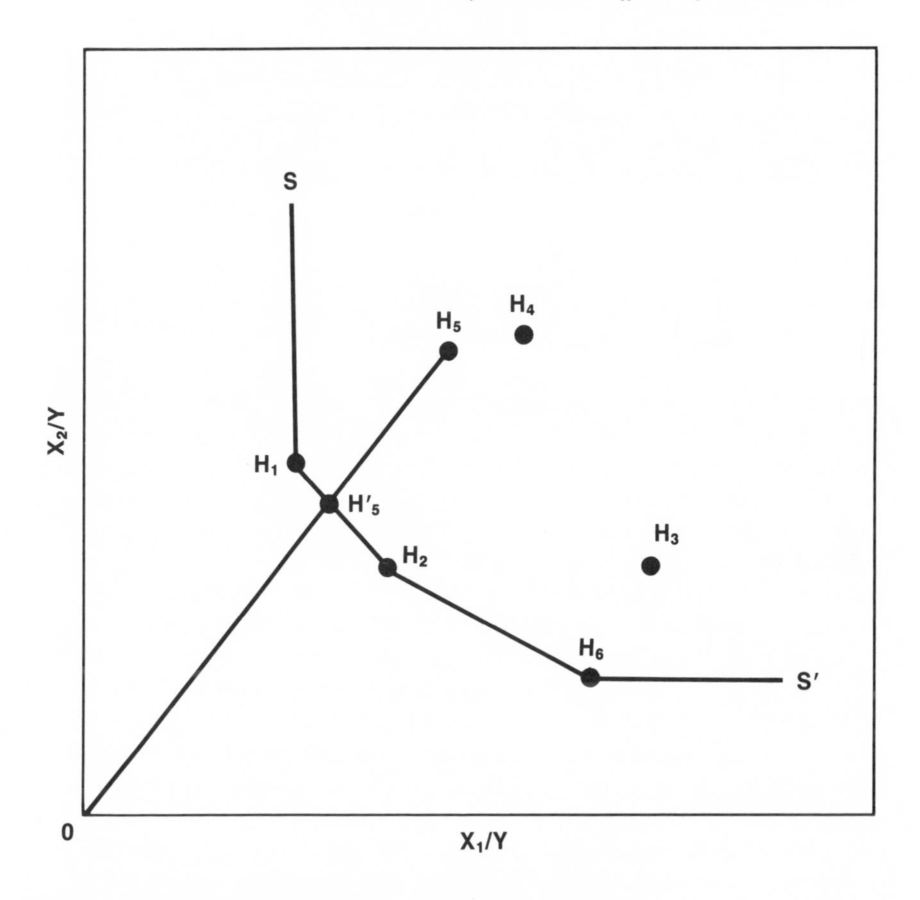


**Fig. 1. An efficient frontier and the relative index of efficiency.**

To measure the efficiency level of the inefficient units, we hold constant the mix of inputs used by these household units. This assumption is equivalent to drawing a line from the origin through the frontier to point $H_5$. The point $H'_5$ on the frontier represents the equiproportional reduction in input per unit output, along each dimension, that will bring unit $H_5$ to the best practice frontier. An index of the relative efficiency of household $H_5$ can be calculated by dividing the length of the line segment $OH'_5$ by the length of segment $OH_5$. The resulting index represents the efficient proportion of inputs needed for technically efficient service provision: $H'_5$ provides the same output as $H_5$ using only a fraction, $OH'_5/OH_5$, as much of each input. This ratio $OH'_5/OH_5$ will range from 0 to

1, with 1 denoting the presence of a unit on the frontier. The larger the index, the more productive the household compared to other households with the same mix of inputs. In this example, we use only two input dimensions and one output. In general the method can accommodate multiple inputs and outputs, but the index will range from 0 to 1, and will always bear the same interpretation.

## The Household Energy Data

Data from the Household Screener Survey (HSS: see U.S. Department of Energy 1981a, 1981b for details) are used to evaluate household service provision efficiency in 1979. The HSS is designed as a follow-up to the National Interim Energy Consumption Survey (NIECS; see U.S. Department of Energy 1980 for details). The aim of the survey is to measure annual energy use and energy expenditures from a national sample of households (excluding Alaska and Hawaii), as well as to collect data on household and dwelling unit characteristics associated with energy use. The data collected include the end uses of fuel by the household, a limited range of energy conservation activities pursued in the previous year, and characteristics of the dwelling unit and household members. Actual energy consumption data are obtained from billing records maintained by the household's fuel supplier(s).

Individual households differ substantially in their level of energy use. Even households with the same physical characteristics may have widely different energy-use patterns because of the multitude of behavioral and structural factors involved. In order to ensure that the differences in household energy efficiency we observe result from general factors, we have directed our attention to the efficiency of categories of households, rather than that of individual households. A typology of households is constructed using HSS data on individual households. Each type of household is defined using the set of eight variables listed in Table 1. Combinations of the values for the eight variables form the set of potential unique household types. As an example, one type of household is located in the Northeast, outside a standard metropolitan statistical area (SMSA), in a weather zone with more than 7,000 heating degree days, and occupies a smaller, single-family, detached dwelling, built between 1950 and 1959, which uses primarily natural gas for house heating, natural gas for water heating, and does not use air-conditioning. Nearly 58,000 unique combinations are possible from the values of the eight variables shown in Table 1.

Households that did not report one of the values listed for each

## Table 1. Variables Used to Construct the Household Typology

| | **Variable Definitions** |
|---|---|
| Census Region: | Four major U.S. census regions as defined by the Census Bureau |
| SMSA Status: | Size of Standard Metropolitan Statistical Area as defined by the Census Bureau in 1970 |
| Weather Zone: | Derived from the seven major weather zones defined by the American Institute of Architects, based on combinations of heating and cooling degree days |
| Year House Built: | Respondent's estimate of the time period when the dwelling unit was constructed |
| Type of Dwelling Unit: | Interviewer's description of the residential structure |
| Main House Heating Fuel: | Respondent's indication of the fuel type used most for space heating |
| Main Water Heating Fuel: | Respondent's indication of the fuel type used most to heat water |
| Air Conditioning Fuel: | Respondent's indication of the fuel type used most for space cooling |

| | **Variable Values** |
|---|---|
| Census Region: | Northeast, Northcentral, South, West |
| SMSA Status: | SMSA >1 million, SMSA <1 million, outside SMSA |
| Weather Zone: | Zone 1--> 7000 HDD and < 2000 CDD, Zone 2-- 5500-6999 HDD and < 2000 CDD, Zones 3-- 4000-5499 HDD and < 2000 CDD, Zones 4 and 5-- < 3999 HDD and < 2000 CDD, Zones 6 and 7-- < 3999 HDD and > 2000 CDD. |
| Year House Built: | Before 1940, 1940-49, 1950-59, 1960-74, after 1974. |
| Type of Dwelling Unit: | Mobile Home, SFD[*] < 6 rms., SFD > 5 rms., SFA[**], MF[†], 2 to 4 units, MF more than 4 units. |
| Main House Heating Fuel: | Natural Gas, LPG, Fuel Oil, Electricity |
| Main Water Heating Fuel: | Natural Gas, LPG, Fuel Oil, Electricity |
| Air Conditioning Fuel: | Not used, Natural Gas, or Electricity |

[*] Single Family Detached

[**] Single Family Attached

[†] Multi-Family

of the eight variables in Table 1 have been dropped from the study. For example, households that do not use one of the four fuels listed in Table 1 as their primary house heating fuel have been deleted. Approximately 1,400 household types are identified in the HSS using the eight-variable typology. Household types that do not represent at least five individual households from the HSS were also deleted from the typology. The resulting 207 household types (hereafter called "units") represent a common set of households in the continental U.S. exhibiting values of the eight characteristics used to select the typology. The household types contain an average of ten individual households from the HSS. The typology represents a "mainstream" set of household types.[2]

The household type is the unit of analysis in this study. Certain characteristics for each household type, such as the number of occupants, income, energy use, etc., are defined using the average of that characteristic for the individual households forming a type. Consider a large single-family detached house built after 1974 outside a large SMSA in weather zone 3 in the West Census Region. This particular household type uses electricity as the primary house and water heating fuel. Five individual households in the HSS share these characteristics. Their average electricity consumption is 121 million Btu with a standard deviation of 24 million Btu. The electricity use for this type of household is the average electricity use for the five individual households. Our focus is on the differences in characteristics between different types of households and not on comparisons between similar households.

In sum, the units in the household typology represent at least five households from the HSS. For each unit, the values for such continuous variables as income, energy costs, etc., are defined as the means of the values for the individual households within each type.

---

[2] The choice of the minimum number of households represented by a household type is arbitrary. Setting the frequency boundary at five represents an attempt to balance the trade-off between the diversity of types and the economy of analysis. The selection of an arbitrary lower boundary raises concern about the representativeness of the household typology. Tests were conducted to determine if the distribution of the values for each of the eight variables differs significantly between the 207 units and the initial household level data from the HSS. The distributions of house heating fuel (chi-square = 9.77, df = 3, $p < 0.01$), water heating fuel (chi-square = 12.12, df = 3, $p < 0.01$), and air conditioning use (chi square = 6.71, df = 1, $P < 0.01$) in the 207 unit typology differ significantly from the distribution of these fuel use characteristics in the HSS. The typology contains more units that use natural gas for house heating and water heating, and more units that use air-conditioning, than expected from the HSS distribution.

**Specification of the Efficiency Measure**

Specifying a model to measure the efficiency with which energy intensive household services are provided involves selecting measures of resource inputs and service outputs. A complete specification of the service provision model should contain measures of both fixed and variable inputs, as well as measures of all services provided in the household. Measures of fixed inputs might include the nature of the building envelope, the appliance stock, and the heating and cooling equipment. Variable input measures might include fuel expenditures or the quantities of fuels used within the household.

A wide range of output indicators may be relevant to the analysis, such as the size of the area within the dwelling, the number of gallons of hot water used for bathing and household cleaning, the temperature of the hot water, the number of occupants, and so on. Unfortunately, the measured energy use of households in the HSS is not partitioned into fixed (nonweather sensitive) and variable (weather sensitive) loads.[3] Therefore, these types of services cannot be individually represented in the analysis, and the specification of service outputs cannot distinguish completely between indicators for base and variable components of energy use.

The specification of the inputs and outputs for the cost-efficiency measure is as follows:

Annual expenditures for electric and nonelectric fuels are used as the measure for the input resources in the model. Using dollars as a measure of resource inputs has several advantages. From the consumer perspective, dollars are the basic measurement units applied in household management decisions (Kempton and Montgomery 1982). From the perspective of the energy analyst, evaluating the household energy budget in dollars, as opposed to the physical units of fuel used, reduces many of the energy aggregation problems discussed by Slesser (1978). The price of a particular fuel reflects the cost to suppliers of converting natural resources, such as coal and crude oil, into usable fuel, such as electricity and

[3] Methods are available to separate base load energy use from house heating energy use using monthly billing data (Fels 1984). An attempt was made to estimate base load energy use from the annual billing and annual climate data available on the HSS, but the results were unsatisfactory. No simple methods exist to separate house cooling energy use from base load energy use that have been verified from household submetered data.

fuel oil. Fuel prices also reflect the relative overall energy cost of the transformation process. Thus, fuel costs (that is, quantity used times the price) are also proxies for the primary energy required to deliver fuel to the consumer.

The service output indicators available from the HSS are indirect measures of service provision: the number of occupants in the household and number of rooms in the dwelling. The number of household members is related to the base load energy use of a household over an annual cycle; that is, the greater the number of occupants, the greater the demand for hot water, etc. The number of rooms in a dwelling unit is a proxy for dwelling unit size. For any given set of climatic conditions, the size of a dwelling unit is one key factor in determining the heating and cooling load of the structure, in conjunction with the thermal properties of the building envelope and the level of heating and cooling required by the occupants.

The above model estimates the efficiency of the physical transformation of resource inputs to service outputs. The efficiency index is a measure of technical efficiency with respect to costs. As a result, our energy-efficiency measure reflects cost efficiency.[4] Baxter et al. (1986) introduce an alternate measure of technical efficiency, where the inputs are specified as the physical quantities of the fuels used within the household.

## Distribution of Cost-Efficiency Scores

A histogram of the efficiency scores obtained from the two-input, two-output household energy services model is presented in Figure 2.[5]

---

[4] Our use of the term cost efficiency should not be confused with the measurement of allocative (price) efficiency. Allocative efficiency requires a change in the proportions among inputs that allocates inputs to their highest valued uses. Allocative efficiency thus introduces the opportunity cost of inputs to the measurement of productive efficiency. See the discussion in Baxter (1984) for more details.

[5] The optimization model used to construct the efficiency frontier is sensitive to outliers. An outlier is any unit with an extreme input-output combination and is a concern only if it is part of the efficiency frontier. Timmer (1971) and Schinnar (1980a) both suggest that outlier units be dropped from the data to reduce the ambiguity of the results. Since no a priori criteria exist for defining units as outliers (as opposed to very efficient units), intervals are established about the mean values of the input and output measures corresponding roughly to plus or minus two standard deviations about each mean. Units with input or output values falling outside the intervals are deleted. The efficiency frontier constructed from the subset of units in the typology is a conservative estimate of the best performance observed in the sample.

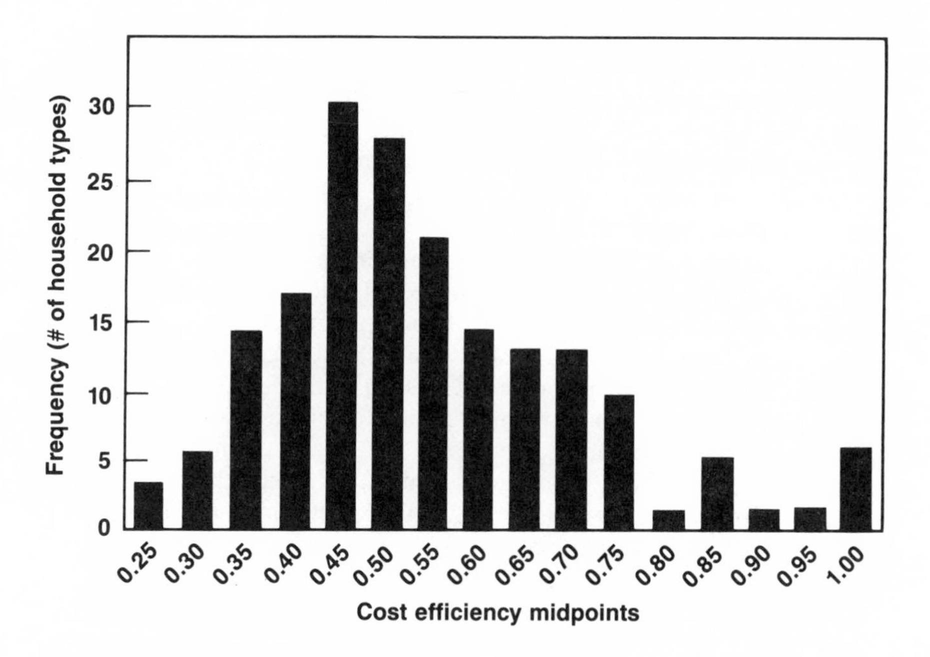


**Fig. 2. Histogram of cost efficiency scores.**

Units with scores of 1.0 define the efficiency frontier. Units with scores less than 1.0 are inefficient: the size of the score indicates the unit's distance from the frontier with smaller scores located farther from the efficiency frontier. The cost-efficiency distribution is skewed markedly to the right. Only about 10 percent of the units have efficiencies greater than 0.75. Most of the units are far from the best practice frontier. The mean cost efficiency score is 0.54 (std. dev. = 0.16, N = 181). A unit on the efficiency frontier using the same mix of inputs as a unit with an efficiency score of 0.54 would use a fraction (0.54) as much of each input for the same level of service.

## DETERMINANTS OF EFFICIENCY AND DEMAND

This section presents a regression analysis of household energy efficiency and compares the results of this analysis to a household energy demand equation.

In order to identify those variables that account for the observed differences in cost efficiency, we regressed the efficiency scores on a host of locational, household, and dwelling variables. This exploratory regression analysis is also a way to identify potential biases in the efficiency measure due to the incompleteness of the HSS data. Including bias indicators in the regression analysis is one way to control confounding influences on the technical efficiency measure.

As an example, consider the potential bias introduced to the measure of technical efficiency by examining base and variable load energy use together. Clearly the greater the heating load placed on a household, all other factors being equal, the larger the variable component relative to the base load component of energy use. The result of the weather sensitive shift from a base load dominant to a variable load dominant household is that the efficiency model specification yields a pattern of technical efficiency scores that are highest in those geographic areas with the smallest combined heating and cooling loads. Introducing indicators of climatic variation in the regression analysis is one way to detect and control for the presence of such regional biases.

The independent variables are grouped into three broad categories, representing factors related primarily to location (such as climate, energy price, and the availability of certain fuels), household characteristics (such as income, education, and the age distribution of the occupants), and the type of dwelling unit (such as single-family or multi-family structures). The first column in Table 2 presents the results of a regression using ordinary least-squares techniques, with cost efficiency as the dependent variable. The table reports standardized regression coefficients. Thus, within the regression model, the relative sizes of the effects of the explanatory variables may be compared. Two of the variables included in our original regression model, the mean price of nonelectric fuels and the occupant-to-room ratio, were dropped due to their collinearity with other explanatory variables.

The results of a regression analysis of cost efficiency indicate that units with higher cost efficiencies in 1979 are located in moderate climates, face lower electricity prices, have higher nonelectric fuel costs per dollar of electric fuel costs, heat houses with electricity, occupy mobile homes or units in larger multi-family dwellings, own the unit they occupy, have lower incomes, and have more children per adult, (with the adults being older and the children younger), other factors in the model held constant. The size of the heating load accounts for the most variation in cost efficiency. In addition, the choice of heating fuel, the size of the cooling load, and the average electricity price all account for substantial variation in cost efficiency.

**Table 2. A Comparison of Two Measures of Energy Use: Total Energy Expenditures Versus Cost Efficiency**

| | Cost-Efficiency Analysis | Expenditures Analysis |
|---|---|---|
| Heating Degree Days (Base 65) | -0.6248** | 0.4968** |
| Cooling Degree Days (Base 65) | -0.3875** | 0.2595** |
| Mean Price of Electricity (1979$/Btu x $10^6$) | -0.3139** | 0.2262** |
| Non-Electric: Electric Fuel Cost Ratio (1979$) | 0.4877** | 0.0537 |
| Dummy: Uses Air Conditioning (1=yes) | -0.1894** | 0.1305* |
| Dummy: LPG Main House Heating Fuel (1=yes) | -0.1111* | 0.0975* |
| Dummy: Fuel Oil Main House Heating Fuel (1=yes) | -0.4122** | 0.3810** |
| Dummy: Elect. Main House Heating Fuel (1=yes) | 0.4075** | 0.0214 |
| Mean Age of Adults (years) | 0.1074* | -0.0747 |
| Mean Age of Children (years) | -0.0968* | 0.0452 |
| Educational Attainment of Respondent (years) | -0.0059 | -0.0499 |
| Household Income in 1977 (0-$50000 by $5000) | -0.1608* | 0.2796** |
| Child to Adult Ratio | 0.2082** | 0.1216* |
| Dummy: Rental Unit (1=yes) | -0.1497** | 0.0482 |
| Dummy: Mobile Home (1=yes) | 0.1281* | -0.1941** |
| Dummy: Single-Family Detached < 6 Rooms (1=yes) | -0.0786 | -0.2146** |
| Dummy: Single-Family Attached (1=yes) | -0.0341 | -0.0650 |
| Dummy: Multi-Family Bldg. 2-4 Units (1=yes) | 0.0197 | -0.2043** |
| Dummy: Multi-Family Bldg. > 4 Units (1=yes) | 0.2414** | -0.4300** |
| Reference Group for Dummy Variables: Single-family detached units with more than five rooms, natural gas main house heating fuel, no air conditioning, occupied by the house owner. | | |
| $R^2$ | 0.7260 | 0.7635 |

* significant at $\alpha = 0.10$

** significant at $\alpha = 0.01$

Note: Coefficients reported are standardized regression coefficients

**Comparing Utilization and Efficiency Analyses**

The above findings about the efficiency implications of heating fuel choice, fuel mix, and rental versus ownership have not been explicitly reported in the literature on national energy-use trends. Uncovering novel relationships may result either from the questions asked (that is, the selection and formulation of variables) or the methods used to address the questions (the efficiency measure used). Would the results differ if, using the same household typology and variables, a regression was performed using total energy expenditures as the dependent variable instead of the efficiency measure estimated in this chapter?

The second column of Table 2 presents the results of a regression using total household energy expenditures (representing an estimate of the total energy use by the household in primary energy equivalents) in 1979 dollars as the dependent variable. Both regression models in Table 2 are estimated using the same independent variables and the same data from the household typology.

A comparison of the significant coefficient estimates between the two models is instructive. Coefficient estimates that are statistically significant in one regression model but not the other include:

- ratio of nonelectric to electric fuel use,
- dummy variable for electricity as the main house heating fuel,
- dummy variable for renters,
- dummy variables for single-family detached and smaller multi-family dwellings, and
- mean ages of adults and children.

Cost efficiency and total energy costs are negatively correlated ($r = -0.71$), so, in general, variables that are significant in both regressions may be expected to have opposite signs. However, the ratio of children to adults is positively associated with both cost efficiency and total energy costs.

Important differences exist between characteristics that identify low expenditure households and those that identify efficient households. Greater use of nonelectric fuels relative to electric fuel, use of electricity for house heating, and ownership of the dwelling are characteristics associated with efficiency but not expenditures. Households with older adults or younger children are also more efficient, but these characteristics cannot be systematically associated with lower expenditures. In addition, all dwelling unit types have lower energy costs than the reference group (except for single-family attached units), but only mobile homes and units in larger multi-family dwellings are significantly more cost efficient.

**Discussion of Results**

The results of an efficiency analysis of household energy use corroborate the findings of earlier studies that focused on the level of energy use in the household.[6] As expected, the price of fuels, the income level, and the type of dwelling are all important factors related to energy efficiency. Climate, though also an important factor statistically, is included in the regression analysis to control for the bias in the efficiency measure due to the presence of both variable and base load energy-use components in the input variables.

The analysis reveals the existence of other factors that also contribute significantly to the efficiency level. These factors, the use of electricity in heating, the ratio of nonelectric to electric fuel inputs, and household tenure status are discussed in more detail below.

Households with electric house heating are surprisingly more cost efficient than the comparison group of units using natural gas for house heating. Several factors may account for the greater efficiency of electrically heated units. First, electrically heated units may have greater proportionate investments in the fixed inputs for service provision (greater levels of insulation relative to the climate, for example). Since the 1960's, U.S. electric utilities have required increased energy-efficiency standards for electrically heated homes in order to slow the rate of capacity expansion caused by growing peak demand requirements. Second, with the high price of electricity, households with electric heating may be more cost conscious than others, and therefore may decrease the level of service in ways not reflected by the service output meas-

---

[6] Most energy-demand studies conclude that the price elasticity of demand is inelastic and negative, while the income elasticity of demand is positive though small (see, for example, Parti and Parti 1980; Scott 1980; Capper and Scott 1982; Garbacz 1983). Other studies conclude, however, that given a stock of appliances, price and income elasticities are zero (see, for example, Wills 1981). Yet other studies examine a broader variety of factors thought to influence household energy use. A U. S. DOE (1981c) study of natural gas and electricity use in single family homes estimates that house-heating energy use with electricity is half that of natural gas, and that the number of occupants is a major factor in the level of hot water use for both electricity and gas. Hirst, Goeltz, and Carney (1982) find that heating use is more important than cooling use in determining the level of total energy use. They also find that the number of adults has a much greater effect on total energy use than does the number of children, that older houses use more energy than newer houses, and that the floor area of a dwelling is an important determinant of energy use.

ures (e.g., by lowering thermostats or heating only a portion of the house) or may increase the use of fuels not measured in the study. Thus, the reader should not conclude from this analysis that electricity is the heating fuel of choice under all circumstances. More detailed efficiency studies would help determine if the energy-efficiency standards of homes using fuels other than electricity for heating should be increased. The results do suggest, however, that policies to encourage or discourage a particular heating technology must be examined in light of both the fixed and variable costs of house heating.

In an apparently contradictory result, the analysis shows that units with higher expenditure ratios for nonelectric relative to electric fuel inputs achieve greater efficiency levels. The "paradox" is explained, however, if we consider that these ratios are zero for all electric units and thus in general have no implications for electrically heated homes. This relation, then, indicates that for homes with a mix of the measured fuels, the variable efficiency of service provision may be raised by shifting a greater relative share of the variable cost to nonelectric fuels—by either reducing electricity use or substituting nonelectric fuels for electricity when possible. Knowledge of the accompanying change in fixed costs is essential to determine if the overall efficiency of service provision increases with either shift. Greater emphasis on the management of electric-fueled services is required, and the economics of fuel substitution in the household should be examined carefully.

As a group, renters are less efficient than owners of larger single-family detached units using natural gas for house heating, a relatively inefficient group of owners. Renters lack incentives to increase efficiency because many of them do not react directly to a known energy price schedule. In a building where the landlord pays for all or part of the energy requirements, the renter faces energy price signals only indirectly, in the form of increased rental fees, reduced flow of capital to building maintenance, or both. As a result, renters are not provided with direct incentives to invest in fixed inputs to increase efficiency, and they are unlikely to modify household management practices to reduce the use of variable inputs when their savings are indirect at best. Similarly, landlords lack incentive to invest in fixed inputs to increase efficiency, since increases in variable costs may be passed to building occupants through greater rental fees. In those cases where the renters pay the fuel bills, the tenant is often unwilling to make capital improvements in the landlord's property.

The three variables of primary interest in the cost-efficiency analysis—use of electric heating, fuel mix, and household tenure status—are not significant in the energy expenditures regression analysis. Thus, our results suggest that the interpretative differences between a conventional energy consumption analysis and an efficiency analysis are not due to a difference in the selection and formulation of the independent variables or the data used. Instead, the results imply that the two measures, utilization and efficiency, reflect different aspects of the energy performance of dwellings and appliances, and the energy-related behavior of the occupants. The efficiency measure contains different information than a measure of total energy use.

## CONCLUSIONS

This study of household service provision efficiency provides new insights into household energy use. The study described above indicates the importance of several factors not considered or found significant in previous studies. The results indicate that electrically heated homes are more efficient than those heated by natural gas, that those homes with higher ratios of nonelectric to electric fuel inputs are more efficient, and that owner-occupied dwellings are more efficient than rental properties.

We are aware of shortcomings to our analysis stemming from the data limitations. For example, the HSS contains no direct measures of service outputs. As a result, the models do not account for some of the variations in the range or level of service provided. Another limitation is the lack of data on fixed inputs. Without such data, the present research is limited to an evaluation of household energy efficiency in relation to variable inputs only.

Given the limitations of the present data, one approach that might reduce the interpretative difficulty of comparing the efficiency of units in different climates involves subdividing the HSS data into climatically similar regions and performing an efficiency analysis of the data within each region. A household production function could be estimated within each region and questions would focus on the patterns between regions. For example, is one type of dwelling unit always more efficient than others in each region?

Three major types of services dominate the energy requirements of most U.S. households: home heating, hot water use, and services requiring appliances. If reasonable proxies are developed for the fixed inputs and service outputs for these major services, then a better estimate of household energy efficiency is possible. These limitations suggest that future residential energy surveys should consider collecting data that enable researchers to evaluate energy use with respect to the services households demand and not just the absolute amount of energy used.

As researchers begin to focus on the collection of more appropriate output indicators, efficiency analysis should be considered a new tool to apply to the design and evaluation of conservation programs. Our results indicate that household characteristics associated with the level of energy use differ in important respects from those characteristics related to efficiency (see also Baxter 1984). Conservation programs may reap greater energy savings, or equivalent savings at lower cost, by focusing on the inefficient users as the target population, particularly if the sources of inefficiency are no less amenable to change than the causes of high energy use. The evaluation of a conservation program's effectiveness would then concentrate on the changes in the efficiency of the target group before and after the program. Program managers and evaluators would be able to measure changes in the level of energy use relative to what that energy is used for, not just the absolute level of energy use.

Finally, frontier analysis provides researchers with a new method to examine the role of occupant behavior in the provision of energy related services. Given the data necessary to specify a relatively complete service provision model, the relative efficiency of households would be determined mainly by the quality of the fixed inputs (building envelope, appliances, etc.) and household management practices. Such a study conducted in a quasi-experimental environment would enable researchers to identify the characteristics of inefficiency at the behavioral level. At present, research in residential energy use is still a long way from explaining the overall effect of occupant behavior on energy use. The application of the productive efficiency approach takes research a step further in this direction—ultimately, to help occupants provide desired household services at the lowest possible cost.

## References

Baxter, L.W. "Measuring the Energy Efficiency of Households: An Application of Frontier Production Function Analysis." Ph.D. dissertation. University of Pennsylvania, 1984.

Baxter, L.W., S. Feldman, A.P. Schinnar, and R. Wirtshafter. "An Efficiency Analysis of Household Energy Use." *Energy Economics* 8:62-73, 1986.

Capper, G. and A. Scott. "The Economics of House Heating; Further Findings." *Energy Economics* 4:134-38, 1982.

Charnes, A., W.W. Cooper, and E. Rhodes. "Measuring the Efficiency of Decision Making Units." *European Journal of Operational Research* 12:429-444, 1978.

Farrell, M.J. "The Measurement of Productive Efficiency." *Journal of the Royal Statistical Society* A 120(3):253-258, 1957.

Fels, M.F. "The Princeton Scorekeeping Method." Center for Energy and Environmental Studies, Report No. 163. Princeton University, Princeton, NJ, 1984.

Garbacz, C. "A Model of Residential Energy Demand Using a National Household Sample." *Energy Economics* 5:124-28, 1983.

Hirst, E., R. Goeltz, and J. Carney. "Residential Energy Use Analysis of Disaggregate Data." *Energy Economics* 4:74-82, 1982.

Kempton, W., and L. Montgomery. "Folk Quantification of Energy." *Energy* 7:817-827, 1982.

Parti, M., and C. Parti. "The Total and Appliance-Specific Conditional Demand for Electricity in the Household Sector." *Bell Journal of Economics* 11:309-21, 1980.

Quigley, J.M. "The Production of Housing Services and the Derived Demand for Residential Energy." *Rand Journal of Economics* 15:555-67, 1984.

Schinnar, A.P. "Measuring Productive Efficiency of Public Service Provision." Fels Discussion Paper No. 143. School of Public and Urban Policy, University of Pennsylvania, 1980a.

Schinnar, A.P. "An Algorithm for Measuring Relative Efficiency." Fels Discussion Paper No. 144. School of Public and Urban Policy, University of Pennsylvania, 1980b.

Scott, A. "The Economics of House Heating." *Energy Economics* 2:130-41, 1980.

Slesser, M. *Energy in the Economy.* New York: St. Martins Press, 1978.

Timmer, C.P. "Using Probabilistic Frontier Production Functions to Measure Technical Efficiency." *Journal of Political Economy* 79:776-94, 1971.

U.S. Department of Energy. *Residential Energy Consumption Survey: Consumption and Expenditures, April 1978 through March 1979.* Energy Information Administration, DOE/EIA-0207/5. Washington, DC, 1980.

U.S. Department of Energy. *Residential Energy Consumption Survey: 1979-1980 Consumption and Expenditures, Part I, National Data.* Energy Information Administration, DOE/EIA-0262/1. Washington, DC 1981a.

U.S. Department of Energy. *Residential Energy Consumption Survey: 1978-1980 Consumption and Expenditures, Part II, Regional Data.* Energy Information Administration, DOE/EIA-0262/2. Washington, DC 1981b.

U.S. Department of Energy. *The National Interim Energy Consumption Survey: Exploring the Variability in Energy Consumption.* Energy Information Administration, DOE/EIA-0272. Washington, DC, 1981c.

Wills, J. "Residential Demand for Electricity." *Energy Economics* 3:249-55, 1981.

# Public Support for Local Government Regulation to Promote Solar Energy

*Barbara J Burt*
*Arizona State University, West Campus*
*Max Neiman*
*University of California at Riverside*

## INTRODUCTION

Although it is a matter of controversy, it is reasonable to claim that government will play a significant role in encouraging or promoting alternative energy sources, including solar energy. Despite the oscillations of public concern regarding energy issues—now high anxiety, now complacency—government at all levels will continue to manage the nation's energy problems, in part by encouraging solar energy. In this connection, local governments have, according to numerous observers of the energy scene, a unique role to play in the nation's energy policy (Burt 1983; Cigler 1982; Morris 1982; Wilbanks 1982). Local governments, it is said, are in a better position to tailor energy policies to the local climate, terrain, geology, or economic base.

While local governments might best be able to judge the suitability or fit of policy to its local circumstances, localities are also most constrained with respect to their options concerning whether to rely on incentive or regulatory approaches to increasing solar energy use. Incentives rely on subsidies, tax credits, and research

*Energy Efficiency: Perspectives on Individual Behavior*

grants designed to enhance a particular energy resource or system. Governments can provide programs to increase the number of individuals capable of installing and maintaining solar energy systems. Deductions can be provided in the tax code allowing reductions in tax bills upon the installation of a solar system. Increasing the number of grants for research into photovoltaics is another example of an incentive approach to encouraging solar energy.

Regulatory approaches elevate a particular objective to a publicly sanctioned norm. Requiring the installation of solar water heaters for swimming pools, building code regulations regarding passive solar design, subdivision regulations governing solar access, and improving landscaping with a view towards maintaining solar energy use are examples of a regulatory approach. Regulatory approaches mandate an action or rule, and if individual behavior is not consistent, then coercion by government sanction is applied. On the other hand, incentive methods increase the rewards and/or lower the risks of adopting a particular set of behaviors.

Different levels of government in the United States vary in the degree of choice they have in selecting from among incentive or regulatory approaches that affect human behavior. It is generally correct to argue that the federal government and the states have more incentive-type options than do localities, and that localities are more inclined to employ regulatory approaches to achieving public ends. Localities have fewer financial resources for incentive approaches, which require direct expenditures to beneficiaries or indirect expenditures through tax benefits. Explicit limitations on fully exploiting existing tax resources (e.g., property tax rate limitations) also limit support for incentive methods. Furthermore, localities have many regulatory devices and long experience in dealing with development controls through the usual planning process. There is, consequently, a tendency to focus on physical controls via development regulation as a means of achieving energy objectives. In short, few localities have incentive-based programs either because they cannot afford them or because they are not authorized to spend money for such purposes.[1]

Localities interested in achieving such energy objectives as encouraging greater use of solar energy are constrained by circumstances. Institutional factors incline localities to take a regulatory approach to energy policy objectives such as encouraging greater use of solar energy. Yet regulatory approaches tend to cause considerable opposition from those people and institutions

[1] Some localities do not have incentive-base programs such as the Municipal Solar Corporation in Oceanside, California.

being regulated. Individuals prefer to qualify for rewards, rather than being the focus of sanctions (Neiman 1980). As localities try to go beyond the housekeeping actions of energy management and try to achieve large-scale changes in energy systems and energy consumption, it is likely that the programs formulated will be regulatory in nature and that they will encounter substantial opposition, for the reasons mentioned above and because of the anti-regulatory temper that currently prevails in the United States.

In early Spring 1982, the California municipality analyzed in this study considered a set of regulatory policies to mandate greater solar energy use. The publicity and widespread community discussion attending the examination of these proposed policies provided a valuable opportunity to assess residents' support for mandating the use of solar energy. This paper explores the relationship between residents' support for *required* use of solar energy, and a variety of factors that theory and extant research suggest as explanations for variation in support for solar energy. This research examines the tension between municipalities that are institutionally inclined to formulate regulatory approaches to public policies, and citizens that resent and are likely to oppose regulatory approaches. The implications for altering energy system selection among households through regulatory processes will be addressed in the conclusion of this chapter.

While the city council was considering a package of solar energy options that might be required of homeowners, a telephone survey was conducted of a random sample of 500 respondents from the study site. The study site is a southern California city of approximately 172,000 residents. The sample was selected using the random-digit dialing method (Landon and Banks 1977; Tyebjee 1979). Of the 500 respondents selected, 471 provided completed interviews. The interview schedule was designed to include questions regarding the socioeconomic status of respondents, their general political affiliations and attitudes, their attitudes about energy problems, and their evaluations of conservation and solar energy programs that were being discussed in the region. A comparison of demographic characteristics of the sample interviewees and the results of a special city census provide support for the representativeness of the study sample.

## STUDY MEASURES AND FRAMEWORK FOR ANALYSIS

In order to assess the support for the city's package of possible requirements for using solar energy, a series of eight statements were employed to correspond with the components of the city government's package. These statements are reported in Table 1. Each respondent was asked to indicate a "yes" or "no" response for the statements. The categories were coded as "0" for a no response and "1" for a yes response. A simple index of support for regulations requiring solar energy was computed by adding the value of each response. The index score ranges from 0-8. Table 1 also reports the distribution of responses (yes and no, with undetermined responses and nonresponses excluded for each item respectively). The analysis excludes the respondents who were renters, since the proposed regulations only influenced them indirectly. The data analysis focuses on those respondents who would have to consider immediate and direct costs to themselves, but including the renters in the analysis does not change the results. In any case, the sample size, with renters excluded from the analysis, was 345. Table 1 also reports the item-index correlation and the Cronbach's Alpha (0.750). The index of support for local government regulations designed to increase the use of solar energy is the dependent variable in this study.

In selecting the explanatory variables in this study, several components of the proposed solar energy regulations are important. First, the proposed policies deal with encouraging solar energy. Consequently, theory and research bearing on support for solar energy generally is relevant. In their literature review regarding public opinion and energy, Farhar, Unseld, Vories, and Crews (1980) report that opinion polls consistently indicate high levels of support for coal, oil, nuclear, and even other unconventional sources such as wind and geothermal. While the review also documents the paucity of study into solar support, they conclude that "Public attitudes toward solar energy can only be described as positive."

The review also indicates that higher socioeconomic status, a commitment to energy conservation, and fears of energy shortages are associated with general support for solar energy. Additionally, these linkages are supported by research regarding environmental protection. The material dealing with public support for environmental protection is pertinent because a major component in the rationale often used to justify public support of solar energy is that it is a renewable, non-depletable resource, and the greater the reliance on solar resources, the more one preserves such resources as coal and oil and the less one must rely on the putatively risky sources, such as nuclear power. In other words, individuals who tend to support environmental protection also tend to support

**Table 1. The Index of Support for Local Government Regulation to Encourage Solar Energy Use**

| | Percentage Yes | Percentage No | Item-Index* Correlation |
|---|---|---|---|
| 1. Assuming that solar water heaters save energy, they should be required in homes at the time of sale. | 11.0 | 89.0 | 0.39 |
| 2. If solar water heaters save energy, then they should be required only in *newly* constructed residences. | 61.4 | 38.6 | 0.64 |
| 3. If solar water heaters paid for themselves within 3 years, then they should be legally required in all residences at the time of sale. | 21.4 | 78.6 | 0.58 |
| 4. Assuming that solar water heaters paid for themselves within 3 years, they should be legally required for *newly* constructed residences only. | 63.8 | 36.2 | 0.76 |
| 5. Swimming pool heaters should be legally required of all residences if their cost is recovered within 3 years. | 32.5 | 67.5 | 0.63 |
| 6. Solar water heaters for swimming pools should be required *only* with newly constructed pools. | 55.4 | 44.6 | 0.70 |
| 7. Utility companies should subsidize loans in order to promote the installation of solar water devices. | 23.5 | 76.5 | 0.43 |
| 8. The city should adopt regulations so that access to sunlight is maintained for purposes of using solar energy. | 47.2 | 52.8 | 0.55 |

Cronbach's Alpha = 0.75

* All item-index correlations significant at 0.001 or greater.

actions to increase the use of solar energy. Some support for this assumption is found in the work of Pierce (1979), who documents a positive relationship between indicators of support for resource conservation and environmentalism. Consequently, by assessing the literature on the determinants of support for environmental protection (Van Liere and Dunlap 1980) it is possible to select variables that can be reasonably posited as determinants of support for promoting the use of solar energy. Although the literature about the correlates of support for environmental protection is often contradictory and erratic, the literature consistently indicates a positive relationship between support for environmental protection and socioeconomic status, when the latter is measured in terms of standard measures like income, occupation, or years of schooling, although the very wealthy do tend to be less supportive than the middle class (Morrison and Dunlap 1980). Based on such findings and on the assumed overlap between influences on support for solar energy use and environmental protection, it seems reasonable to select indicators of socioeconomic status as possible explanatory variables for the index of support for city regulations to promote greater solar energy use. This approach is supported by one of the few studies that directly probes the determinants of support for resource conservation (including alternative energy sources); this study finds socioeconomic status positively related to a composite index of resource conservation (Honnold and Nelson 1979).

Another variable that has begun to persistently emerge as a significant explanatory measure of support for environmental protection is partisan identification with a particular political party or ideology. This factor is quite important, since for many years environmental protection and resource conservation have been viewed as consensus issues, cutting across ideological and partisan lines (Buttel and Flinn 1978; McEvoy 1972). However, research has hinted at partisan and ideological cleavages emerging around a variety of specific environmental issues. For example, recent research has indicated that Democrats appear to be more supportive than Republicans of environmental protection (Constantini and Hanf 1972; Dunlap and Gale 1974) and resource conservation (Honnold and Nelson 1980; Mazmanian and Sabatier 1981). Since partisanship is associated with support for environmental protection, it might also reasonably be expected to be associated with support for government regulations designed to prod greater use of solar energy.

But the proposed solar energy regulations are not likely to be evaluated by citizens solely in terms of their relevance to the *solar-*

ness of the issue. In this study, one is not merely asking the citizen to express support or opposition to solar energy as such. One is also gauging the response to *regulation* as a method used by government to direct the behavior of the jurisdiction's residents and property owners. Consequently, it is necessary to integrate the fact that specific actions comprising the index of support for promoting solar energy also ask the respondent to swallow a substantial degree of implied government intervention into previously private matters. There is considerable literature indicating that partisanship, ideology, and attitudes about government's role in the economy are important in affecting a person's willingness to accept government intervention and regulation. Consequently, partisanship is also expected to be related to support for regulation generally, with Democrats more receptive than Republicans. Conservatives are expected to be less willing than liberals to accept government regulation. And individuals that generally believe in keeping government out of management of the economy are going to oppose intervention through regulation to achieve greater use of solar energy (Monroe 1975; Erikson et al. 1980; Flanigan and Zingale 1979).

Based on the previous discussion, one can develop a schema to guide the selection of variables that can yield fruitful explorations of differences in support levels for local government regulations designed to increase the use of solar energy. Of course, standard background variables should be included based on the assumption that such measures are at least crude indicators of past socialization, self-interest, and inclinations towards certain attitudes relevant to policy. Among the most important and frequently used background measures are those of social status. This study employs the following background measures: age, minority status, and three measures of socioeconomic status (years of schooling, annual family income, and size of the residential structure). In addition, the following measures of political predispositions are used: partisan identification and a self-rated political philosophy scale. This analysis also uses a number of indicators of attitudes relevant to policy. These include indices of support for environmental protection and support for government intervention in the management of the economy.

## ANALYSIS

### Bivariate Results

The findings indicate that several background variables are

related to support for mandating solar energy use.[2] These variables are age, schooling, and minority status. The age of the respondents is inversely related to support for solar energy regulation. The relationship is moderate and apparently due to the strong linear trend among regulation opponents, the results indicate that the proportion of opponents rises from 29.3 to 53.3 percent between the 35 or younger and 60 or older categories. The reverse of the pattern occurs among the supporters. Schooling also appears to have a moderate relationship among regulation opponents, where the proportion of opponents appears to generally decline as the years of schooling increases. Among regulation supporters, the proportion of supporters is greatest in the highest education category, although the pattern is not as consistent as among opponents of solar energy regulations. Minority status is relatively strongly related to support for solar energy, with 54.5 percent of the monority respondents in the supporter category and only 22.6 percent of the nonminority respondents in this category. Forty percent of the nonminority respondents are found in the opponent category.

In addition to the background measures, all of the political and policy-relevant measures of attitudes are related to support for solar energy regulations. Among supporters of environmental protection measures by government, 41.2 percent are supporters of mandated solar energy use, while among opponents of environmental protection, 52.8 percent oppose such regulations. Partisan identification is weakly, but significantly, related to solar energy regulation. Opponents tend to be more Republican, and as one moves towards the strong Democratic category, the proportion of supporters of solar energy regulation increases. Stronger patterns of a similar nature are evidenced for the measures of political philosophy and support for government intervention in the economy. It is apparent that as the respondents become more conservative, support for solar energy regulation declines markedly. Finally, as support for intervention in the economy by government increases, the proportion of regulation for solar energy also increases.

Overall, the bivariate results suggest that the older, non-minority, less-educated respondents are *less* supportive of solar energy regulation. Democrats, political liberals, and supporters of environmental protection and economic intervention by government are more likely to support local government regulations promoting solar energy use. The significance of these patterns is that solar energy regulation seems to be related to how one feels about government

[2] Detailed bivariate results are available from the authors.

regulation as a means of affecting and shaping the behavior of energy customers. No matter how highly regarded solar energy is, the attitude of this study's respondents is clearly modified when solar energy is sent to the citizen as a regulatory package.

**Multivariate Results**

In order to assess the relative and collective importance of the independent variables in explaining support for solar energy, and to explore the relationships that might be masked in the bivariate analysis, we now turn to a multivariate assessment of the data. Standard least-squares regression and discriminant-function analysis were performed on the data (Aldrich and Cnudde 1975; Klecka 1980). The results were the same, whether regression analysis was employed on the raw score of the solar energy regulation index or whether discriminant-function analysis was used on the collapsed categories of the solar energy index. The discriminant-function analysis provides a more satisfactory interpretation of differences in support for solar energy regulation, for reasons explained below.

In applying the discriminant-function analysis, the trichotomized version of the solar energy regulation index is used initially. The most important features of discriminant function analysis for the purposes of this study relate to its ability to assess the relative importance of particular variables in differentiating between the supporters and opponents of locally mandated solar energy regulations. In this study, respondents are assigned, on the basis of their discriminant scores, to the three categories of support regarding local government regulation to promote the use of solar energy (that is, supporters, undetermined respondents, and opponents). These classifications are then compared to the actual category into which the respondent falls. Using these classification results, one can draw some conclusions concerning the "power" of the independent variables to collectively predict the value of the dependent variables, which in this study are the three categories of support for local government solar energy regulations. In addition, discriminant-function analysis provides coefficients for each of the independent variables. These coefficients are interpretable as indicators of the relative importance of individual independent variables.

Table 2 reports the results of a stepwise discriminant-function analysis, in which independent variables are included in the analysis only if they contribute a significant amount of discriminating power (capacity to discern differences in supporter, undetermined, and opponent function scores). The classification results indicate substantial power to accurately classify respondents into the actual categories for the supporters and opponents categories. Among the undetermined group, however, the results indicate that the functions do rather poorly. Among those respondents in the opponents category, 50 percent are correctly classified, and among the supporters category, about 49 percent are correctly classified. This is well above the 33 percent correct classifications that one might expect by chance alone. Among the respondents in the undetermined category, only 40 percent are correctly classified. The overall power of the discriminant functions is significant, but modest. This result is due to the large number of incorrect classifications among the respondents in the undetermined category. The analysis suggests that for the purpose of clarifying the relative importance of the independent variables in differentiating between supporters and opponents of solar energy regulation, it is useful to eliminate the undetermined group from the analysis.

Table 2 reports the results of a two-group discriminant-function analysis (stepwise), with the undetermined group not included in the analysis. The findings indicate that of the 226 supporters and opponents classified, nearly 70 percent were correctly classified. Moreover, 65.0 of the 132 opponents were correctly classified. A larger percentage of the supporters, approximately 73 percent, were correctly classified. The usefulness of the discriminant function analysis is underscored by the significance of the final Wilks' Lambda (Significance level = 0.0000) and by the moderate canonical correlations coefficient (0.656).

The enumeration of the standardized-function coefficients, indicating the relative importance of the independent variables in contributing to the discriminating power of the function, suggests some differences in the relative importance of the independent variables. Only six of the independent variables survive the stepwise analysis. Minority status stands out as the most important variable, followed by income, partisanship, the index of support for environmental protection, schooling, and political philosophy respectively. The multivariate results, in short, affirm the bivariate findings, which indicate that the more affluent, better educated, minority, politically liberal, Democratic, environmentalists support solar energy regulations.

## CONCLUSION

The relatively small proportion of respondents that fall into the

**Table 2. Discriminant Function Analysis. Support for Solar Energy Regulation Index is the Dependent Variable.**

| Analysis: Three Groups — Independent Variable | Function Coefficient | Analysis: Two Groups — Independent Variable | Function Coefficient |
|---|---|---|---|
| Minority Status | 0.724 | Minority Status | 0.724 |
| Family Income | 0.309 | Family Income | 0.309 |
| Partisanship | -0.306 | Partisanship | -0.306 |
| Support for the Environment | -0.268 | Support for the Environment | -0.268 |
| Political Philosophy | 0.268 | Political Philosophy | 0.268 |
| Schooling | 0.205 | Schooling | 0.205 |

**Function Characteristics**

| | Function 1 | Function 2 | | |
|---|---|---|---|---|
| Eigenvalue | 0.376 | 0.021 | Eigenvalue | 0.376 |
| Percent Variance | 94.6 | 5.4 | Percent Variance | 100.0 |
| Canonical Correlation | 0.523 | 0.145 | Canonical Correlation | 0.656 |
| Significance of function | 0.0000 | 0.6990 | Significance of function | 0.0000 |

**Classification Results**

| | Percent Correctly Classified | | Percent Correctly Classified |
|---|---|---|---|
| Supporters (N=94) | 48.9 | Supporters (N=94) | 72.7 |
| Opponents (N=132 | 50.0 | Opponents (N=132) | 65.0 |
| Undetermined (N=119) | 40.3 | | |
| Overall 46.4 percent correctly classified. | | Overall 69.5 percent correctly classified. | |

supporter category for regulations designed to promote solar energy might appear discouraging to those who are interested in hastening the diffusion of solar energy use. The data suggest that the group supporting a regulatory approach is quite unusual, reflecting the oft-discussed more affluent, better educated, and ideologically liberal segment of the population, joined by politically liberal minority residents.[3] This group comprises a relatively narrow seg-

[3] Just over 91 percent of the minority residents identified themselves as Democrats, and 55 percent said they were ideologically liberal.

ment of the population; the number of respondents categorized in this study as supporters of regulatory approaches to increasing the use of solar energy is approximately one third less than the opponents (132 versus 94). Given the temper that currently prevails regarding the general issue of regulation and the almost automatic skepticism that greets any proposal for new government initiatives in the area of regulation, one might find the results reported here disheartening.

This study's findings demonstrate the influence that social values and political attitudes have in the support for local government regulations promoting solar energy. The variables most important in predicting support or opposition for promotion of solar energy are a mix of background characteristics and social or political attitudes. In the bivariate analysis, however, the attitudinal indicators have a consistent relationship with the solar energy index. If one were to assume that background characteristics are temporally prior and are influential in the formation of social and political attitudes, then the attitudinal correlates become more important in explaining the opposition to government regulation of the use of solar energy. For example, 54.5 percent of the minority respondents fall in the support group of the solar energy index. Further investigation of these respondents' political affiliation shows that more than 90 percent identify themselves as Democrats. Democrats, as we argued earlier, tend to be more supportive of government activism and of regulation than Republicans. Surely, minority status is temporally prior to formation of partisan loyalty, but minority group identification was no doubt important in the formation of an affiliation with a political party.

Localities, as mentioned above, are limited because of their institutional structure in their choices about how to assume an active role in adopting and implementing energy policy. It appears that tensions generated by attitudes of opposition towards regulation and support for solar energy programs can be a serious constraint in the promulgation of widespread solar energy use through regulation. A host of attitudinal mechanisms are triggered that for many citizens transform the perceived objective of policy from one of increased use of solar energy to one that has as its main focus the extension of yet another governmental tentacle into individual's lives.

Energy policy decision makers, however, may take note that individual solar regulations may be acceptable even while a complete package may not. Requiring installation of solar hot water heaters and swimming pool heaters in *newly* constructed residences is supported by a majority of the study respondents. Evidently, the public is willing to impose additional regulation on buyers of new residences, but opposes regulations that would apply to existing dwellings. This finding opens the regulatory process to initial steps, and as solar energy becomes more accepted it may be possible to increase the number of ways solar can be required in residences and commercial structures.

The findings here, in short, contribute to the growing body of evidence that issues such as the degree of government's role in the protection of environmental amenities and the promotion of resource conservation through the use of nondepletable resources (e.g., the sun) are evaluated on the basis of social and political values. Whether the relevance of the ideological division is reflective of the current tenor of the times, which is clearly anti-governmental and anti-regulatory, or whether this division represents an enduring feature of energy policy generally and solar energy in particular remains to be seen.

**References**

Aldrich, J. and C. Cnudde. "Probing the Bounds of Conventional Wisdom: A Comparison of Regression, Probit, and Discriminant Function Analysis." *American Journal of Political Science* 35:578-608, 1975.

Burt, B.J. "The Political Dimension of Conservation Policy." Paper prepared for the Annual Meeting of the Western Political Science Association, Seattle, Washington, March 25-27, 1983.

Buttel, F.H. and W.L. Flinn. "Social Class and Mass Environmental Beliefs: A Reconstruction." *Environment and Behavior* 10:433-450, 1978.

Cigler, B.A. "Intergovernmental Roles in Local Energy Conservation: A Research Frontier." *Policy Studies Review* 1(4):761-776, 1982.

Constantini, E. and K. Hanf. "Environmental Concern at Lake Tahoe: A Study of Elite Perceptions, Backgrounds and Attitudes." *Environment and Behavior* 4:209-242, 1972.

Dunlap, E. and P. Gale. "Party Membership and Environmental Politics: A Legislative Roll-Call Analysis." *Social Science Quarterly* 55:670-690, 1974.

Erikson, R.S., N.R. Luttbeg, and K.L. Tedin. *American Public Opinion: Its Origins, Content, and Impact.* New York: John

Wiley and Sons, 1980.

Farhar, B.C., C.T. Unseld, R. Vories, and R. Crews. "Public Opinion About Energy." *Annual Review of Energy* 5:141-172, 1980.

Flanigan, W.H. and N.H. Zingale. *Political Behavior of the American Electorate.* Boston: Allyn and Bacon, 1979.

Honnold, J.A. and L.D. Nelson. "Support for Resource Conservation: A Prediction Model." *Social Problems* 27:220-233, 1979.

Klecka, W.R. *Discriminant Analysis.* Beverly Hills, CA: Sage Publications, 1980.

Landon, E. and K. Banks. "Relative Efficiency and Bias of Plus-One Telephone Sampling." *Journal of Marketing Research* 15:294-299, 1977.

Mazmanian, D. and P. Sabatier. "Liberalism, Environmentalism, and Partisanship in Public Policy-Making: The California Coastal Commission." *Environment and Behavior* 13:361-384, 1981.

McEvoy, J. III. "The American Concern with the Environment." In *Social Behavior, Natural Resources, and the Environment*, ed. W.R. Burch, 214-236. New York: Harper and Row, 1972.

Monroe, A.D. *Public Opinion in America.* New York: Dodd and Mead, 1975

Morris, D. *Self-Reliant Cities: Energy and the Transformation of Urban America.* San Francisco: Sierra Club Books, 1982.

Morrison, D.E. and R.E. Dunlap "Elitism, Equity, and Environmentalism." Paper prepared for the Environmental Policy Seminar at the Annual Meeting of the American Sociological Association, New York, 1980.

Neiman, M. "Zoning Policy, Income Clustering, and Suburban Change." *Social Science Quarterly* 61:666-675, 1980.

Pierce, J.C. "Water Resource Preservation, Personal Values, and Public Support." *Environment and Behavior* 11:147-161, 1979.

Tyebjee, T. "Telephone Survey Methods: The State of the Art." *Journal of Marketing* 43:68-78, 1979.

Van Liere, K.D. and R.E. Dunlap. "The Social Basis of Environmental Concern: A Review of Hypotheses, Explanations and Empirical Evidence." *Public Opinion Quarterly* 44:181-197, 1980.

Wilbanks, T.J. "Local Energy Initiatives and Consensus in Energy Policy." Paper prepared for the Committee on Behavioral and Social Aspects of Energy Consumption and Production, National Research Council, March 1982.

# A Decade of Residential Energy Research: Some New Directions and Some Speculations About the Future

*Bonnie Maas Morrison*
*Michigan State University*

## INTRODUCTION

In October 1983, the Family Energy Research Team of the Institute for Family and Child Study sponsored a working conference, "Families and Energy: Coping with Uncertainty," which was held at Michigan State University on the tenth anniversary of the Arab Oil Embargo. The conference was the first national attempt to focus on residential energy research from both a technical and a behavioral perspective. This conference brought together leading national and international residential energy researchers, who not only gave state-of-the-art research papers, but also shared their expertise in establishing a national agenda for future residential energy research.

Residential energy research has generally been conducted by two different sets of researchers—engineers and social scientists. Each group works quite independently of the other. It has commonly been assumed by members of both groups that the determinants of energy use and/or conservation could be explained by either the physical characteristics of a house (including weather) or by the characteristics of the occupants. However, in limiting the research to one set of measures to the exclusion of the others, researchers

*Energy Efficiency: Perspectives on Individual Behavior*

could only partially explain the variation in residential energy use.

In the often quoted Twin Rivers Study, for example, Socolow (1978) found a two-to-one difference in energy consumption in 28 carefully monitored identical townhouses. These results led to the conclusion that "unpredictable behavior patterns of the occupants introduce a large source of uncertainty" (Socolow 1978). Others have also come to a similar conclusion (Sonderegger 1978; Welch et al. 1982; Lipschutz 1983). In this case the physical scientist identified the need for social research.

The social scientist, on the other hand, has not generally attended to the physical side of the residential energy picture (including the measurement of energy use), except to describe the types of dwelling units, the number of doors, windows, and appliance packages, etc. This approach has led to basic measures of household characteristics, attitudes and *self-reported* energy related behaviors, without serious consideration of the physical aspects of a home or its energy consumption—(Aronson and O'Leary 1983; Becker 1978; Craig and McCann 1978; Darley 1978; Farhar et al. 1979; Leonard-Barton 1981; Olsen 1981; Stern and Kirkpatrick 1977; among others).

The Family Energy Research Team, however, from the inception of its work in 1974, was convinced that there was more to energy use and consumption changes than a careful assessment of the residence or the characteristics of the people could provide. Taking an ecological or systems approach to the study of energy consumption in households allowed analysts to incorporate the sociophysical properties of energy use into their analyses (Morrison 1976; Morrison and Gladhart 1976 and 1978; Gladhart 1977; Gladhart et al. in press).

The family energy research team was also convinced that interaction between physical and social scientists would have positive long-range effects on future residential energy research. The conference was therefore planned to stimulate conversation and interaction between the two scientific groups, so that future residential energy research would be better informed by both—this time working together. To this end, four keynote addresses (Daniel Yergin, David Snider, Maxine Savitz, and Elliot Aronson), set the stage for the continued importance of residential energy research, while two plenary session papers put the residential energy research into first, an international perspective (Lee Shipper) and second, a social science/behavioral perspective (Paul Stern). Fifty research papers were also presented that addressed such broad issues as: measurement, methods, evaluation of audits and educational programs, equity issues and theory.

The essence of this chapter is: (1) To put into perspective the outcome of the conference and to report the most important insights gained from it, and (2) to indicate some of the future issues residential energy researchers need to address.

Therefore, in keeping with the theme of the 1984 ACEEE Summer Study, "Doing Better: Setting An Agenda for the Second Decade," I will share with you what was presented at the working conference. I will not review the work contained in each paper, but instead highlight some of the most promising contributions presented, while indicating some of the shortcomings that exist. Finally, I will address the general areas where important work remains.

## NEW DIRECTIONS

The quality of the conference papers showed that the level of research sophistication had gone beyond the early surveys and opinionnaires so prevalent in the energy crisis atmosphere of 1973-74. For example, the research presented at the conferences shows evidence of: (1) greater refinement of measurements of energy uses, and conservation, especially for natural gas and electricity, (2) research methodologies that draw upon several disciplines, making multimethodologies and multidimensional research approaches possible, (3) the use of theory, either in the development of research strategies or for explanation.

### Measurement

Several papers presented at the conference addressed crucial energy-use measurement issues. Most of these measurements were seen as important, because they were more precise than results obtained from previous research. Although not all of these measures have met the test of validity and/or reliability, they are nonetheless worthy of mention.

**Modeling of Residential Natural Gas Consumption.** Taking to task the predictive value of the natural gas model proposed by Lehman and Warren (1978), Harris, Jager, and Zuiches (1984), developed a state-level annual (calendar year) model of residential natural gas consumption for 48 states. This "new" model comes closer to predicting and/or verifying several contemporaneous studies done nationally and locally than did the Lehman and Warren Model, which showed no conservation for the years 1974-76. The basic points of difference between the Lehman and Warren model and the Harris et al. model are the use or nonuse of a "lag parameter" and the use of an annualized state-level natural gas data base compared to a quarterly model from four gas-consuming regions (See Appendix A).

Testing the Harris et al. model of State-level, Annual Natural Gas Consumption (using the American Gas Association data of yearly consumption per household by each state and two customer categories), gives a more accurate reflection of the pattern of residential natural gas conservation that had taken place between 1974 and 1976 than the Lehman and Warren model allowed.

**Measuring Performance of Residential Conservation Programs.** Most evaluations of energy conservation programs, whether utility based or government sponsored, depend on measures of consumer self-reports of conservation actions (attitude, intentions, and reported behavior) via a survey of the various programs (Hirst 1984). This type of evaluation leaves open to question the real effectiveness of such programs and any precise evaluation of the energy saved by the programs. In other words, few evaluation studies come down to an energy-savings analysis of the program. Hirst concludes, therefore, that major strategic planning "rests on a very narrow-and inadequate- data base." Hirst uses two examples "to demonstrate the feasibility and relevance of using utility bills to determine the energy savings attributed to conservation programs." The examples are from the Minnesota Home Energy Audit and Loan Programs (MECS and PUPIP) of the Northern State Power Company (NSP), and the Pacific Northwest Zero-interest Loan Program of the Bonneville Power Administration (BPA).

One of the major contributions of Hirst's paper is the attribution of reduction in use based on participation or nonparticipation in energy audits, and audits plus loan programs. These refined measurements of natural gas use take into consideration: (1) individual households' non-weather-sensitive (baseload) consumption and weather-sensitive (primarily space heating) consumption, (2) total normalized annual consumption[1] for each of three years, (3) influence of house size, number of occupants, degree of insulation,

[1] Normalized annual consumption is an estimate energy use under "typical" weather conditions.

efficiency of heating equipment and thermostat settings, (4) audit and nonaudit program participation, and (5) audit plus loan program participation.

Hirst's analysis assumes that variation in Normalized Annual Consumption (NAC) is a function solely of cross-sectional (i.e., individual household) factors such as demographic and dwelling-unit characteristics, including income, number of household members, floor area of home, age of dwelling units, participation in audit programs, and further, accounts for possible biases in estimating program savings introduced by self-selection into the audit programs.

The major methods used for the NSP and the BPA programs are as follows:

For NSP Programs: (1) the development of a simple model of each household using normalized annual consumption, (2) use of the NAC model results as the dependent variable in a cross-sectional analysis (See Appendix B),

And for BPA Programs: (1) both one and two above, plus (2) an explicit analysis of the decision to retrofit and to participate in the BPA Program.

Using three mathematical models to estimate the savings of natural gas that could be directly attributed to utility conservation programs, Hirst accounts for a small but statistically significant natural gas savings of four percent for MECS participants compared to nonparticipants. A 13 percent reduction is seen for the PUCIP Audit and loan households compared to none for the PUCIP audits only. The results are similar for the BPA Program as well.

The findings suggest that careful attention to analysis of utility billing data (actual consumption data) for pre- and post-program participants, including comparison with non-participants, is a sound basis for determining program energy saving. Using self-reports of attitudes, intentions and reported behaviors for program evaluation was seen as an insufficient basis for judging the effectiveness of conservation programs.

**Longitudinal Measures of Residential Energy Consumption and Conservation.** The Family Energy Study's initial energy project was a major study that analyzed residential energy consumption and conservation on the same sample for a period of five years (between 1973-78). This study included the measurement of natural gas, electricity and fuel oil data (depending on household energy mix) controlled for degree days and translated to Btus (Gladhart 1984). Both consumption and conservation were measured as dependent variables. In the *regression model of consumption,* the independent variables used were: dwelling unit variables (including size of heating and cooling load, number of windows, number of doors, number of rooms heated and air-conditioned), and lifestyle variables (including annual family income, family size and employment status).

Although the $R^2$ for consumption was 0.49 and 0.55 there was an overall high degree of uniformity and stability of the various coefficients and their t-values from year to year. Also, the coefficients for a majority of the variables had a high level of statistical significance.

In the *regression model of conservation,* several variables representing lifestyle were used (including income and family size, family life cycle, ratio of annual energy costs to per capita family income, and when appropriate, "reported" attempts to conserve). Dwelling unit measurements were initially used but did not improve the fit of the equations; therefore only reported conservation actions were used to explain conservation. These included: installed ceiling insulation, wall insulation or clock thermostat (within the previous two years), and an index of reported conservation behaviors, along with reported installation of ceiling, wall and crawl space insulation between spring and fall of 1977, and a report of receiving personalized weatherization analysis, and participation or nonparticipation in an infrared experiment (Zuiches 1978). The results of the regression analysis of conservation for 95 households with complete data for each period of the five years indicates that from 20 to 25 percent of the variation was explained in the period 1973-76 ($R^2$ = 0.16), whereas 59 to 69 percent of the variation between 1975-78 ($R^2$ = 0.77) is explained.

The analysis showed that using survey and utility billing data does allow a preliminary examination of the determinants of energy consumption. Over a five year period of examination, the data were uniform, stable, and statistically significant. Also, the use of utility billing data and reported conservation measures and behaviors allows an understanding of the determinants of energy conservation, which indicates that families "conserve incrementally" (Gladhart 1984). Thus, when consumption and/or conservation are measured over a period of time, it is possible to show that more families undertook conservation as their sensitivity to energy price rises increased.

**Separating Lower Thermostat Settings from Extra Insulation.** Using just billing and weather data, Fels and Goldberg (1984), have measured the separate effects of lower thermostat settings and extra insulation, and their results indicate that thermostat setbacks have played a greater role in natural gas conservation than have either structural changes or modifications in appliance usage, especially for the first four years following the Arab Oil Embargo (1973). More recently, conservation of natural gas can be attributed to reductions in water heater temperatures and appliance usage as well. In both cases the effects of retrofitting remains small.

Using aggregate utility sales data (1970-1982) for nearly a million households in New Jersey that use natural gas for heating, a conservation index resulting from analysis of normalized annual consumption indicates a dramatic 26 percent drop in natural gas consumption from the peak year of 1973 to 1982. Eighteen percent of the 26 percent reduction was realized in the first four years after the embargo.

Fels and Goldberg were able to analyze the effects by exploring three physical parameters in three nonoverlapping four year periods. The physical parameters were: (1) base level consumption, used independently from outside temperatures, (2) the heating slope, representing the constant amount of fuel required per drop in degree day (outside of houses), and (3) break-even or reference temperature corresponding to the outside temperature when no heating fuel is required (See Appendix C).

Because the standard error in the three physical components is relatively large (30 to 40 percent) compared to the standard error for the total consumption index $\Gamma$ (only 3 percent), the results in making any comparison of the components should be considered "necessarily tentative." However, they are nonetheless interesting, especially when ($\tau$) is associated with lowered thermostat setting, ($\beta$) is associated with structural retrofits or furnace efficiency improvements, and ($\alpha$) is associated with a decrease in appliance usage or increased appliance efficiency. The findings indicated that the reduced natural gas consumption was due to reduced interior temperatures (1970-73); since that time the reductions in natural gas consumption can be attributed to reduced consumption by appliances. The implication, although tentative, is that this measurement (given validity and reliability testing) could help in an understanding of where in the residence conservation is occurring.

**Multimethodological Approaches**

In the ten year search for more precise and comprehensive measurements of energy consumption and consumption changes, methodological "experiments" have been tried. Most recently these experiments have included both technical (nonbehavioral) measures and behavioral measures. Evidence is beginning to suggest that when both types of measures are used, a more accurate picture emerges of the determinants of energy consumption and changes in predicted energy use. The following examples illustrate this point.

**Behavioral Model.** Larry Lewis, market researcher at Consumer's Power Company in Michigan, has attempted to identify energy consumption determinants, monitor specific energy conservation efforts, and predict energy-use changes. Like many other analysts, Lewis used nonbehavioral econometric models or physical/structural models that depict energy consumption as a function of customer appliance stock, demographics, socioeconomics, and reported conservation factors, as well as housing characteristics, energy price increases and rates. These methods, though commonly used and conceptually sound, only explain between 30 to 50 percent of the variation in energy usage.

The behavioral model developed by Lewis, although not proposed on the basis of social science concepts or theory, is, however, a measure that recognizes that: (1) energy consumption is conceptually nothing more than the aggregation of the energy required to operate each of the customers' appliances, and (2) appliance energy usage is a function of frequency of use, duration and level of use (taking into consideration size, age and efficiency).

Not able to undertake expensive submetering of appliances to correctly measure frequency of use, duration or level of use, Lewis conducted an "experimental survey" to determine the feasibility of incorporating appliance usage behavior questions in residential questionnaires. The responses to reported appliance usage behavior are then integrated into nonbehavioral energy models to improve understanding of factors affecting the level of energy use.

The behavioral energy model was used in a time-of-day study to ascertain the effects of electrical rate structures on off-peak and on-peak usage. Data on appliance ownership, usage (frequency), demographic and socioeconomic characteristics were compared with actual metered usage. The behavioral energy model predicts that 43.51 percent of usage is on-peak, compared with 42.17 percent for the actual metering. Although the method of behavioral energy-use modeling is in the early stages of development, it appears promising.

**Sociophysical Determinants of Residential Energy Use.** In an example of multimethod, multidimensional residential energy research, Cramer and his colleagues have attempted to integrate social/behavioral variables and physical/structural variables into a single causal model of household energy use. The *social variables* included: household size, income, and expected future energy prices. These were considered measures of household demand and economic constraint variables. Other measures of *social variables* included attitudes, values and access to information. The *physical variables* used were the engineering determinants of energy use, including: inventory of major appliances, size of house, self-reported behaviors regarding frequency of use of appliances, and the presence of air-conditioning.

The theoretical causal model as articulated by Cramer et al. is: (1) energy use is explained as a function of physical variables entirely, and (2) the physical variables are in turn explained by the social variables.

In this type of model, social variables do not directly change energy consumption, but are indirectly related to energy use due to their links with the physical variables (Cramer et al. 1984). This causal model was tested on single-family dwellings, for summer electricity use, in a community where air-conditioning accounted for a significant part of the electrical use. The sample included 192 households.

The data were analyzed with multiple regression techniques using three different equations. The first equation, basically the engineering model (only physical variables included), explained 51.3 percent of the variance in summer electricity use. In the second equation, when only the social variables were regressed on summer electricity use, 33.5 percent of the variance of summer electricity use was explained. The third equation included all the engineering and social variables. This equation explained 58.4 percent of the variance of summer electricity use, adding 7.1 percentage points to the explanatory power over the use of engineering variables alone. Although not a huge improvement in explanatory power, this example shows that both physical and social variables are related to energy use in a simple causal chain; however, the three equations do not indicate how the social variables *indirectly* affect summer electricity use (Cramer et al. 1984). In further analysis, family income and number of persons in the household appeared to be the most important indirect determinants of energy use. They are significantly related to the appliance index (P = 0.01), and in the case of income, significantly related to size of house as well. Improved measures were called for, including decision-making and management consideration of differing lifestyles (Cramer et al. 1984).

**Monitoring Techniques**

The problems of explaining residential energy use have perplexed most researchers because of the large proportion of unexplained variance found. Although on $R^2$ of 80 percent was reported by Cramer et al. (1984b) in the study of structural-behavioral determinants of summer electricity uses when different housing types were combined in a regression equation (apartments and townhouses), it is rare to see an $R^2$ greater than 60 percent. Reported frequency of use of various appliances is often the only measure of behavior (Lewis 1984; Cramer et al. 1984, a and b).

In order to overcome this lack of explanatory power, researchers have most recently undertaken research that uses multidisciplinary approaches and/or multimethodologies that employ sophisticated monitoring devices, while attempting to pay attention to the physical structural aspects of the house and the behavior or lifestyle characteristic of the occupants.

**Garrett House.** Turner et al. (1984) are investigating the outcome of various energy conservation strategies on: (1) the actual energy consumption and (2) the lifestyle changes resulting from the conservation strategies.

The research team is multidisciplinary, with a home economist acting as full time coordinator of all activities, including intensive work with the households; a sociologist, who is responsible for the attitudinal and behavioral data collection and analysis; and a mechanical engineer, who is responsible for the monitoring instrumentation and analysis of the energy data base.

One of the several unique aspects of this study is that the housing stock remains constant, while the families change (live-in "experimental families", spend either six weeks in the winter months or four weeks in the summer months). A baseline study establishing family norms (where *no* energy-conserving strategies are introduced) is obtained in the first week of each family's residency in Garrett house. This baseline study establishes both energy consumption norms and general household activities. After the first week, various conservation strategies are initiated, either alone or in combination with other strategies. At the beginning of each "new strategy week" a list of instructions and information is given to the family, and, when appropriate, demonstrations are also given in the use of various conservation techniques or technologies.

A prequestionnaire (prior to residency) is used to ascertain pre-Garrett House energy consumption behaviors and attitudes, as well as demographic information. During the residency, daily logs of activities are recorded indicating time, place and duration of the various activities. A weekly questionnaire concerning the strategies employed during a one week period elicits attitudes about comfort, convenience and demands on time and attention. A post-questionnaire on attitudes is completed immediately at the end of each residency by both family adults. A follow-up questionnaire is administered a year later.

The energy data is recorded by a microcomputer capable of monitoring 48 channels (37 channels are actually in use). These channels are scanned every 10 seconds. The data collected includes room temperature, humidity, hot water consumption, electrical consumption and run times of several appliances. The computer also monitors exterior temperature, humidity, and wind speed, while measuring solar radiation. The microcomputer data, gathered along with manual daily readings of natural gas and total water consumption, is compared with predicted energy savings. This study pays close attention to behavior and lifestyle as well as physical measures of energy consumption. "Over time, the comparison of the lifestyle dimensions across the various strategies will increase knowledge of energy consumption and related behavior of families," in spite of the nonrandomly selected small sample (Turner et al. 1984).

**Integrated Engineering and Social Science.** A second example of a multidisciplinary (engineering and social science) multimethodological study is the work being conducted by Kempton and his associates at Michigan State University, reported in the paper by Weihl, Kempton and DuPage (1984). Although the paper is mainly a report of the instrument package developed "to record the behavior which affects space heating and water heating energy consumption," the project itself is an attempt to integrate the data from the instrumentation with data gathered using ethnographic and observational techniques. The impetus for this study was the same frustration indicated by many other researchers--the inability to explain variation in energy use based on engineering analyses or survey research of attitudes and demographic characteristics in isolation.

This study measures "inferred behavior" rather than actual behavior. Several dependent variables such as furnace on-time, window open-time, door open-time and vent on-time are used to infer various energy management behaviors, while thermostat settings, exterior and several interior temperatures are also being monitored. Hot water heater on-time is a second dependent variable. By recording time of day, duration and volume of flow at individual hot water use points, inferred "hot-water-use events" of major categories of hot water uses will be distinguished. Energy-consumption data for electric gas and water are recorded from utility meters.

The analysis (in process) will seek to measure household behaviors and separate the causes of the behaviors into three components: unconscious habits, deliberate management, and deliberate selection for convenience or comfort. The detailed analysis provided on a sample of eight different families in eight houses should improve "understanding of how behaviors and beliefs interact with the residence and its energy systems" (Weihl et al. 1984).

**Theory**

Theory has not been widely used in residential energy research. Since the goal of most of the research has been to describe the energy-use phenomena and to use the facts gained in formulating policy and programs, theory has been neglected. Thus, the opportunity to use theory in providing a rationale needed to account for the phenomena under investigation, or as a framework for conceptualizing and/or implementing research strategies, has often been bypassed or overlooked. This omission has left the results of the investigations with little meaning beyond themselves; in other words, the findings are not translatable into cultural, social or personal implications. Social science and behavioral theory give meaning to the facts and numbers in human terms, much as physical science theory lends explanatory and predictive power to physical phenomena. Although the conference papers exhibited detailed mathematical theoretical models (already discussed in the measurement section), only a few of the papers used or discussed theory from the social science perspective. Of these, only two will be discussed here.

Diffusion of Innovation Theory. Using a combination of diffusion of innovation theory (Rogers 1983), energy diffusion theory (Darley and Beniger 1981) and communication network theory (Rogers and Kincaid 1981), Yearns designed the research of Home Energy Audits using theory to structure her research strategy and analysis.

Based on the theoretical characteristics of "earlier adopters of innovation", Yearns "hypothesized that the implementation of energy-conserving innovations is positively associated with income, education, size of home, information-seeking, and connectedness, and inversely associated with age and attitudes about the legitimacy of the energy crisis" (Yearns 1984). Although the multiple regression analysis did not identify participation in the Home Energy Audit Program as a significant predictor of implementation of conservation improvements,

> *The analysis did reveal that the most significant characteristics of earlier adaptors of energy-conserving innovations were age, connectedness, and problems with the current energy situation. Younger households, households that are more connected with others in the social system, and, households that are experiencing problems with the current energy situation, are more likely to make structural changes to improve the energy-efficiency of their homes* (Yearns 1984).

**Folk Theory.** In order to make sense of the world, people develop their own theories about how things work. Kempton (1984) in his study of Home Heat Control, explores two folk theories people use to understand and adjust their thermostats--the feedback theory and the valve theory.

According to the *feedback theory,* the thermostat turns the furnace on or off according to room temperature.

> *When the room is too cold, the thermostat turns the furnace on. Then, when the room is warm enough, it turns the furnace off. The setting, controlled by a movable dial or lever, determines the on-off temperature. Since the theory posits that the furnace runs at a single constant speed, the thermostat can control the amount of heating only by the length of time the furnace is on* (Kempton 1984).

This folk theory is most closely related to conventional engineering wisdom of how a thermostat operates as a self-regulating device. The *valve theory,* on the other hand, contradicts conventional technical wisdom about thermostat operation but nonetheless is probably used by many individuals in adjusting their thermostats. In the valve theory, analogies to the automobile gas pedal or a water faucet are often used (i.e., the more the valve is pressed or opened the more speed or water or in this case heat will come forth.) Since the function of a gas pedal and a water faucet is within the everyday experience of most individuals, it is not surprising to find that some people use this theory in the everyday operation of a thermostat.

Kempton postulates that depending on which of these two theories are used, differences in thermostat behavior will occur. Using data from records of thermostat setting collected by Princeton University's Center for Energy and Environment (hourly thermostat setting for a two year period) two homes with differing patterns of thermostat use were used to illustrate that house thermostat adjustments appear to be consistent with the folk theory. In the first case, thermostat settings are the same for several days running, with some minor changes in setting being observed in the morning, at noon and between 5 to 8 pm. This example seems to demonstrate the feedback theory; that is, the thermostat setting is changed only when occupant activities require it, but not at other times.

The second illustration, from data on a second house, shows thermostat changes on an almost hourly basis, with the only period of no change occurring between 1 am and 7 am (during sleep times). The thermostat settings were not done at regular times, but seemed to be random adjustments done on a frequent basis. The authors suggest that the household operating this thermostat adheres to the valve theory (the more the valve is turned up, the faster the furnace runs, the more heat is delivered).

The major point of this interesting and unique analysis is that technology is not always understood by the lay person in the same way that the expert or engineer understands it or desires it to be understood. Therefore, its intended use or operation may be different than its actual use or operation in a household setting.

## SOME SPECULATIONS

From a selected overview of some of the conference papers, it is possible to illustrate some promising new directions being taken in residential energy research. The examples used in this paper were selected on the basis of clearly identifiable new directions in measurement, in methodology, and with the use of theory. Rather than review what these papers have already illustrated, I will suggest some of the shortcomings that exist and then some of the research areas that still need to be addressed.

Using actual energy consumption data in all forms of residential energy research is critical to explanation, evaluation and/or description. However, we all know that energy data itself is expensive and difficult to manipulate. There is no perfect mathematical model that is useful for all energy research or for all energy uses. There is no perfect model that addresses the need to measure baseload energy compared to space heating energy compared to appliance usage, while also addressing the physical properties of a place, the human behavior within it or the weather that surrounds it. Several of the papers given at the conference advance towards this goal, but still leave 20 to 35 percent of the variance in energy consumption or conservation unexplained. Why is this? Is it the inconsistency in the variables measured? Is it that the variables have been poorly measured? Is it that the "best" variables have not been identified? It is that we have used mostly "convenience measures" (i.e., measures that already exist in some data base)? Is it that theory has not been used to frame research questions; is it that we have not learned from each other, and therefore are unaware or ignore the strengths that exist in the knowledge base?

The answer to all these questions is an unequivocal yes. However, not all is ill in residential energy research. Present research is tackling some of these knotty residential research problems. Future research must also address these critical areas to overcome past shortcomings.

From within the conference a working group identified four major emphasis areas that are needed in future residential energy research. I will quote from the working group's draft report *Families and Energy: Ten Years After the Oil Embargo* (Zuiches and Savitz 1984).

*The Research Agenda:*

1. a synthesis and accumulation of known results;
2. addressing the fundamental questions of measurement of energy use and conservation;
3. targeting selected substantive research issues, e.g., diffusion and adoption analysis, market segmentation analysis, and studies of social and corporate institutions affecting energy behaviors;

4. suggesting mechanisms for adding to the knowledge base by small-scale pilot projects, comparative studies, and evaluation studies of on-going programs.

Each of these research areas addresses the central theme of evaluating and learning from past efforts, and each emphasizes the social, economic, and behavioral aspects of energy use patterns.

Quoting from *Energy Use: The Human Dimension* (Stern and Aronson 1984) to further reinforce these ideas:

> *In the past, analysts have frequently been surprised when well engineered energy technologies fail to work as expected, or when carefully planned policies or programs are greeted with public apathy or opposition, or when energy users behave very differently than was predicted or expected. Often the surprise is traceable to the fact that the analysts have not paid enough attention to crucial processes in individual, organization or social institutions.*

Thus analysts need to be sensitive to the human, as well as the physical dimensions of energy consumption in the next decade of residential energy research.

**Acknowledgement**

*This research synthesis was supported by the Michigan Agricultural Experiment Station, Project 3152, located in the Institute for Family and Child Study, College of Human Ecology, Michigan State University. Bonnie Maas Morrison is a professor and assistant dean for Urban Affairs Programs.*

**References**

**Conference Papers from:** Morrison, B.M., and W. Kempton, *Families and Energy: Coping with Uncertainty.* Conference Proceedings. Institute for Family and Child Study, College of Human Ecology, May 1984.

Aronson, E. "Residential Energy Conservation: A Social-Psychological Perspective"

Cramer, J., E. Vine, P. Craig, T. Dietz, B. Hackett, D. Kowalczyk and M. Levine. "Structural-Behavioral Determinants of Residential Energy Use: Summer Electricity Use in Davis." (1984a).

Cramer, J., T. Deitz, N. Miller, P. Craig, B. Hackett, D. Kowalczyk, M. Levine, and E. Vine. "The Determinants of Residential Energy Use: A Physical-Social Causal Model of Summer Electricity Use." (1984b).

Fels, M. and M. Goldberg. "With Just Billing and Weather Data, Can One Separate Lower Thermostat Settings From Extra Insulation?"

Gladhart, P. "Interactions of Prices and Household Characteristics in The Determination of Residential Energy Conservation."

Hackett, B. "Energy Billing Systems and The Social Control of Energy Use in a California Apartment Complex."

Harris, C., D. Jager and J. Zuiches. "Modeling Trends in Residential Natural Gas Consumption."

Hirst, Eric. "Measuring Performance of Residential Conservation Programs."

Hutton, R. and D. McNeill. "Energy Conservation and Vulnerable Groups: Identifying Market Strategies for Home Energy Audits."

Katz, E. and S. Morgan. "The Financing of Energy Conservation Services to Low-Income Households: Alternatives to Grants."

Kempton, W. "Two Theories Used for Home Heat Control."

Kushler, M. and J. Saul. "Energy Conservation and the Low-Income Sector: Assessing Needs, Examining Alternative Programs."

Lewis, L. "The Case for Behavioral Energy Modeling."

Savitz, M. "Ten Years Out – A Problem Not a Crisis."

Schipper, L. "Household Energy Use in Nine Countries: How Much Has Been Saved? Why?"

Snyder, D. "The Energy Outlook For America in the Second Post-OPEC Decade."

Stern, P. "Energy and Behavior: What Have We Learned?"

Turner, C., D. Klett and F. Ahmed. "Effect of Home Energy Conservation Strategies on Lifestyle: Project Overview."

Weihl, J., W. Kempton and D. DuPage. "An Instrument Package for Measuring Household Energy Management."

Yearns, M. "The Home Energy Audit: Its Effect on Conservation Behavior."

**Other References:**

Aronson E., and M. O'Leary. "The Relative Effectiveness of Models and Prompts on Energy Conservation: A Field Experiment in a Shower Room." *Journal of Environmental Systems,* 12, 1983.

Becker, L.J. "The Joint Effects of Feedback and Goal Setting on Performance: A Field Study of Residential Energy Conservation." *Journal of Applied Psychology,* 63, 1978.

Darley, J.M. and J.R. Beniger. "Diffusion of Energy-Conserving Innovations." *Journal of Social Issues* (37) 1981.

Farhar, B.C., C.T. Unseld, R. Vories, and R. Crews. "Public Opinion About Energy." *Annual Review of Energy,* (5) 1980.

Gladhart, Peter M., "Energy Conservation and Lifestyle: A Integrated Approach." *Journal of Consumer Studies and Home Economics,* 1(4) 1977.

Lanoue, R.. *Broadening the Market for Conservation.* Technical Development Corporation, 1981.

Lehman, R.L., and H.E. Warren. "Residential Natural Gas Consumption: Evidence that Conservation Efforts Have Failed." *Science*, 199, 1978.

Leonard-Barton, D. "Voluntary Simplicity Lifestyles and Energy Conservation." *Journal of Consumer Research,* (8) 1981.

Lipschutz, R.D. "Energy Use Patterns in a Large, Multi-Family Building." Energy and Resources Group, University of California Energy Buildings Program, Applied Science Division. Lawrence Berkeley Laboratory, Berkeley, CA, 1983.

Morrison, B.M. "Residential Energy Consumption: Socio-Physical Determinants of Energy Use." *Research and Innovation in the Building Regulatory Process.* P.W. Cook Ed., NBS Special Publication 473, U.S. Department of Commerce/National Bureau of Standards, Washington, D.C., 1977.

Morrison, B.M., and P.M. Gladhart. "Energy and Families: The Crisis and the Response." *The Journal of Home Economics,* 68(1) 1976.

Morrison, B.M. and B. Long, "Energy and Families: The Crisis and the Response." *The Journal of Home Economics,* 70(5) 1979.

Newman, D.K. and D. Day. *The American Energy Consumer.* Cambridge, MA: Ballinger, 1975.

Olsen, M.E.. "Consumers' Attitudes Toward Energy Conservation." *Journal of Social Issues,* 37(2) 1981.

Perlman, R. and R.L. Warren. *Families in the Energy Crisis.* Cambridge, MA: Ballinger Publishing Co., 1977.

Rogers, E.M. *Diffusion of Innovations.* 3rd Ed. New York: The Free Press, 1983.

Rogers, E.M. and D.L. Kincaid. *Communication Networks: Toward a New Paradigm for Research.* New York: The Free Press, 1981.

Socolow, R.H., Ed. *Saving Energy in the Home: Princeton's Experiments at Twin Rivers.* Cambridge, MA: Ballnger Publishing Co., 1978.

Sonderegger, R.C. "Movers and Stayers: The Resident's Contribution to Variation Across Houses in Energy Consumption for Space Heating." In Socolow, Robert H. Ed., *Saving Energy in the Home: Princeton's Experiments at Twin Rivers.* Cambridge, MA: Ballinger Publishing Co., 1978.

Stern, P.C. and E. Aronson, Eds. *Energy Use: The Human Dimension.* New York: W.H. Freeman Co., 1984.

Stern, P.C., and E.M. Kirkpatrick. "Energy Behavior." *Environment* 19(9), 1977.

Welch, L.W., C.J.P. Delange and M.J. Wooldrige. "An Examination of Household Energy Data Collected for Comparison with the CSIRO Low Energy House." Technical Report TRI, CSRIO, Highett, Victoria, Australia: Division of Energy, 1982.

Zuiches, J.J. and M. Savitz. "Families and Energy: Ten Years After the Oil Embargo." Draft Report of Research Needs Working Group, 1984.

## APPENDIX A

**A Lag Parameter Model** (Lehman and Warren, 1978):

$$S = aC + bxH(dZ) + bHZ' \quad (1)$$

where

| | | |
|---|---|---|
| $S$ | = | total residential natural gas consumption |
| $a+b$ | = | are parameters to be estimated |
| $C$ | = | total number of residential energy customers |
| $H$ | = | residential house heating customers |
| $Z$ | = | heating degree days in present calendar year |
| $x$ | = | the lag parameter |
| $Z'$ | = | the number of heating degree days in the three month period lagging a calendar quarter by one month |
| dZ | = | $Z - Z'$ |

**"New" State Level Annual Natural Gas Consumption Model** (Harris et al. 1984):

$$S_{i,j} = aC_{i,j} + bH_{i,j}Z_{i,j} \quad (2)$$

where

| | | |
|---|---|---|
| $S_{i,j}$ | = | total yearly natural gas consumption for 48 states, excluding Alaska and Hawaii |
| i | = | from one to nine years |
| j | = | one to 48 states |
| a | = | coefficient estimated for average nonheating consumption per customer |
| b | = | coefficient estimated for average consumption per heating customer/degree day |
| C | = | total number of residential natural gas customers |
| H | = | residential, house heating customers |
| Z | = | heating degree days |

## APPENDIX B

**Expanded equation:**

$$E_{it} = 3 \quad (1)$$

where

| | | |
|---|---|---|
| $E_{it}$ | = | energy use for household$_i$ in *month$_t$* |
| $a_{ij}$ | = | a coefficient that reflects a household's natural gas use for nonspace heating purposes for household$_i$ and year$_j$ (j = 1, 2, 3) |
| $b_{ij}$ | = | a coefficient of individual household sensitivity to changes in heating degree days (HDD) |
| $HDD_{itj}$ | = | heating degree days for household$_i$ in month$_t$ in year$_j$ |
| $D_j$ | = | a binary variable (dummy variable) equal to one in year$_j$ or zero otherwise |

**Normalized Annual Consumption Equation:**

$$NAC_{ij} + 365*_{ij} + b_{ij}HDD_i \quad j = 1, 2, 3 \quad (2)$$

where

| | | |
|---|---|---|
| $NAC_{ij}$ | = | normalized annual household natural gas consumption for household$_i$ and year$_j$ |
| $365*_{ij}$ | = | coefficient of households use of natural gas for nonspace heating purpose for 365 days of each of three years (j = 1, 2, 3) |
| $b_{ij}$ | = | a coefficient that includes the influence of house size, number of occupants, degree of insulation, efficiency of heating equipment, and thermostat setting, with $b_{i1}$ being weather sensitive for first year and $b_{i2}$ and $b_{i3}$ are weather sensitive for years 2 and 3 |
| $HDD_i$ | = | heating degree days for household$_i$ |

**The Cross-Sectional Equation:**

$$NAC_{ij} = a_o + \frac{\Sigma}{k} a_k Z_{ik} \quad (3)$$

where

| | | |
|---|---|---|
| $Z_{ik}$ | = | a vector of k demographic and dwelling unit characteristics |
| $a_k$ | = | the coefficient of reflecting how weather-adjusted energy consumption depends on the cross-sectional characteristics |

## APPENDIX C

**Fels and Goldman equations** (Aggregate Model, 1984):

$$G_i = \alpha + \beta \, WH_i \, (\tau) \qquad (1)$$

where

| | | |
|---|---|---|
| $G_i$ | = | average daily consumption for *month*$_i$ computed from aggregate utility sales of natural gas |
| $\alpha$ | = | amount of fuel used independent of outside temperatures |
| $\beta$ | = | the heating slope, represents the constant amount of fuel required per degree drop in outside temperature |
| $\tau$ | = | break-even or reference temperature, outside temperature when *no* heating is required |
| $WH_i\tau$ | = | a weighted average of daily heating degree days to base $\tau$ during *months*$_i$ and *month*$_{i-1}$. |

The results of Equation 1 are used to calculate a weather-adjusted conservation index called Normalized Annual Consumption, $\Gamma$.

$$\Gamma = 365 \, \alpha + \beta \, H_o(\tau) \qquad (2)$$

where

| | | |
|---|---|---|
| $\Gamma$ | = | Normal Annualized Consumption |
| $H_{o\tau}$ | = | average annual heating degree-days to base $\tau$ |
| $\alpha$ & $\beta$ | = | same as Equation 1 |

Breaking analysis into *components of change* for three physical parameters ($\alpha$, $\beta$ & $\tau$), using a first-order expansion applied to Equation 2 above.

$$2\Gamma_j - \Gamma_i = 365(\alpha_j - \alpha_i) + H_o(\tau_k) \;(\beta_j - \beta_i) + \beta_k + [H_o(\tau_i) - H_o(\tau_i)] \quad (3)$$

(*base level*) (*shell*) (*temperature*)

where

| | | |
|---|---|---|
| Γj - Γi | = | annual consumption index for three nonoverlapping periods minus the following period j-i (example: 1970-73 minus 1975-78 1975-78 minus 1979-82) |
| $k$ | = | expansion term, found to be independent of the three components ($\alpha$, $\beta$, $\tau$). |

# SECTION IV:

# Home Energy Management

# SECTION IV: Home Energy Management

*Willett Kempton*
*Princeton University*

## INTRODUCTION

The chapters in this section restrict their studies to small samples, while examining behavior at a highly detailed level of analysis. Through ethnographic interviews and direct measurements, they describe the behavior of individuals as they manage home energy use as well as the beliefs and rationales for that behavior.

Erickson makes a unique contribution to this volume by comparing individual energy behavior in two countries. She compares a community in Minnesota with a similar community in Sweden. Sweden has frequently been compared to the U.S., but the differences in energy consumption have too often been explained with unproven stereotypes—for example, contrasting the carefully conserving Swede with the energy glutton American. (These stereotypes, she found, were also held by members of both communities.)

Erickson's data include utility-metered energy consumption, intensive interviews, questionnaires, and daily activity reports. In comparing energy-using activities, she found mixed results. The American community had more of some energy-using behaviors and less of others. From her extensive tabulations Erickson concludes that the stereotypes are neither confirmed nor reversed, and that since the energy effects of behavior seem approximately equal, the lower Swedish overall energy use is likely due to more efficient Swedish houses and appliances.

As an attempt to explain the behavioral differences, she asked her informants why they made certain choices for energy-intensive behaviors. In the American community, they gave as reasons saving time, offsetting stresses of everyday life, and compensating for feelings of powerlessness. By contrast, in the Swedish community, the strongest factors were social relations and expectations (for example, not wanting guests to feel cold), and the need to counteract the long, cold, dark winters.

When asked about factors promoting conservation, the Americans mentioned their agricultural heritage and a negative attitude toward waste. The Swedish community also mentioned their own fuel bills, but unlike the Americans they extended this concern to the national and world economy as well. Conservation was also encouraged by the Swedish moralistic view of society, the love of nature, and the concept of moderation or appropriateness denoted by the Swedish word "lagom".

In the first of three chapters from a project studying home energy management via instrumentation and interviews, Kempton examines hot water use. His sample includes few homes, but those cases are measured in great detail. For example, hot water volume and temperature are measured at each tap in the house, sampled each minute throughout the year. Thus, Kempton disaggregates the energy consumption of the water heater, down to the individual behaviors that determine the quantities of hot water drawn. Interviews with the residents provide complementary verbal data used for interpreting and checking the instrument data.

Kempton tabulates hot water use both in liters and in megajoules of heat contained in the water. By either measure, hot water consumption varies widely across the study houses, and is not explained by obvious factors such as the number of residents. Although he does not provide a quantitative explanation, Kempton illustrates some factors contributing to this variation through description of individual cases. For example, bathing, the highest-volume use, is compared for two individuals, one of whom uses five times as much energy as the other to take a shower.

Kempton calculates efficiency of the hot water piping system by comparing water temperatures at the tank outlet and the taps. Piping system efficiency is completely different from the more commonly computed efficiency of the water heater itself. Efficiency of the distribution system is dependent on occupant behavior. For example, many short uses will be less efficient than one long draw of the same volume (other factors being equal) because the hot water will stand in the pipe and lose heat between short uses. Kempton finds that bathroom sink uses are inefficient, but since total volume is comprised principally of large-volume events, which are over 90% efficient, overall system efficiency is not too bad.

In the second chapter from this data set, Kempton and Krabacher analyze thermostat settings. They find that even today a significant proportion of households maintain high interior temperatures and never practice night setback. Kempton and Krabacher demonstrate a phenomenon that has not previously received much (if any) acknowledgment in the energy conservation literature: in most houses, fall and spring thermostat adjustments are more erratic and swing through wider ranges than in the winter. The authors investigate the transition from the erratic fall pattern to the regular winter pattern by examining individual days during a six-week transitional period at one house.

Kempton and Krabacher propose several new methods for quantitatively describing aspects of a household's thermostat management, including adjustments per day, and mean degree change per adjustment. These numerical measures, along with graphs, illustrate the diversity in thermostat setting patterns in this sample. As in the Kempton hot water chapter, interview data are used to help illustrate the possible reasons for this diversity of thermostat management patterns. Finally, Kempton and Krabacher compare the thermostat settings reported in the interview with those recorded on their instruments. Since they have several reasons to expect participants in their study to be more accurate than those in a typical survey, their analysis can only provide a lower bound to how much surveys might underestimate actual thermostat settings.

Weihl draws from the same data set as the two earlier chapters, but focuses on a comparison of reported schedules with the timing of energy-using behavior. He uses the daily cycle for analysis because he feels it exerts the strongest influence on behavior (in comparison with weekly or monthly cycles). Weihl examines four houses in detail to illustrate differing daily cycles, which he calls "patterns". In each house, he goes into many specifics in detailed comparisons of the reported daily cycle with the instrument-measured thermostat setting, and with water used for bathing, kitchen use, and sinks.

In his summary section, Weihl draws out regularities from these data. Thermostat setting is highly determined by family characteristics. Families with fixed and simple schedules are more likely to have thermostat settings with strong daily cycles. This phenomenon is also demonstrated by the weekend thermostat patterns, which were generally less regular and shifted toward later hours. Weihl points out that different types of families may have very different needs for, and acceptance of, timer-based conservation devices such as clock thermostats. Compared with thermostat cycles, hot water use cycles were less predictable from interview-elicited daily schedules. Bathing was the most schedule-determined of the water uses, while bathroom sink use was the least.

# Household Energy Use in Sweden and Minnesota: Individual Behavior in Cultural Context

*Rita J. Erickson*
*University of Minnesota*

## ENERGY DEMAND AND THE QUESTION OF "LIFESTYLE"

In the late 1970s, Sweden occupied a central place in comparisons of U.S. energy demand with that of other countries. Attention was drawn to Sweden in large part because of studies carried out on Swedish energy use (Schipper and Lichtenberg 1976; Hambraeus and Stillesjo 1977). From these reports came the tantalizing finding that Swedes used only 60% as much energy, per capita, as Americans. How was this remarkable efficiency achieved while maintaining the notoriously high Swedish standard of living?

A central question became the relative importance of behavior versus technology in determining Sweden's lower residential energy consumption–estimated to be 70% to 85% that of the U.S., per capita (Doernberg 1975). Did Swedes use less energy because their daily lives were markedly different from those of Americans, or were Sweden's superbly constructed houses and appliances primarily responsible for its lower consumption?

Researchers lacked the information necessary to answer this question. Although they had access to large-scale, technical

*Energy Efficiency: Perspectives on Individual Behavior*

energy-flow analyses, the aggregated data produced by such studies were inadequate to answer more specific questions, especially with regard to residential patterns. Studies comparing energy consumption among households revealed dramatic individual variation (Darley et al. 1979; Goldstein 1979; Lundstrom 1980; Gaunt and Berggren 1983). Yet this variation could not be explained by the aggregate of technical variables measured, and not enough was known about domestic activities to improve the quality of investigation.

A critical gap existed in knowledge of everyday household patterns and their contexts, as well as the perceptions, attitudes, and values of the residents. Too often, differences in energy consumption were ascribed to differences in "lifestyle." This now-ubiquitous term was neither defined nor documented, however, and stereotypes were invoked to explain differences in energy consumption. Swedes were depicted as being careful energy conservers, living simply, and doing tasks by hand. Americans, in contrast, were thoughtless energy gluttons, living on a lavish scale, and addicted to machines.

To what degree are these stereotypes based in reality? How much of Sweden's lower household energy consumption is explained by differences in Swedish and American "lifestyles?" What are the respective social and cultural forces influencing the decisions of individual consumers?

To address these questions, I conducted anthropological fieldwork for six months in each of two towns: Foley, Minnesota (1980 and 1981) and Munka Ljungby, Skane, Sweden (1982). The single-family dwelling was the unit of research. A mixture of quantitative and qualitative data was gathered from a variety of sources, among them community-wide questionnaires, billing records of utility companies, and interviews with key informants, such as educators, government officials, and merchants.

In addition, a group of "core" households, numbering twenty-two in Foley and twenty-one in Munka Ljungby, participated more intensively in the research. These households completed additional questionnaires and submitted to lengthy interviews. Sociologist Michael Sobel's (1981) operational definition of "lifestyle" as "expressive behaviors that are observable" was adopted and applied in a detailed analysis of key energy-using activities of the core households. Each household member aged six or older kept daily records of engagement in nineteen specified activities during four different weeks.

The discussion that follows presents the reported indoor and water temperatures and some related factors for the Foley and Munka communities. Twelve of the energy-using household activities recorded by the core groups are contrasted for frequency and duration, and are weighed against prevailing national stereotypes. Following this quantitative review are comparisons of social and cultural factors that promote energy consumption, and those that promote energy conservation, in Sweden and in Minnesota. Finally, the sources of discrepancies between attitude and behavior in Foley and Munka are explored.

## DAILY HOUSEHOLD PATTERNS

Based on comparative fuel consumption data gathered for one measurement year, Munka Ljungby core households used, on the average, 86% of the household and hot water fuel that Foley households did. This lower Swedish use was expected, although the percentage is higher than that reported by most studies or estimates. The factors behind this lesser consumption will be discussed shortly.

The relationship between the two communities is reversed for space heating fuel consumed. Here, after degree-day adjustment, Foleyites used 86% of the fuel, on the average, that Munka Ljungbyans did. This unanticipated finding can be explained by physical differences in the housing and by various consumer choices, some of which are presented below.

### Indoor and Water Temperatures

Two weighty choices householders make in relation to total household energy consumption are choices of indoor air and hot water temperatures. Munka Ljungbyans belie the Swedish stereotype in these areas, choosing higher temperatures than do the Foleyites. The average indoor daytime temperature reported by the Munka heating fuel sample was 68 degrees F, one degree higher than the Foley sample average of 67 degrees.[1] Munka's nighttime average temperature, 66 degrees, was two degrees higher than Foley's 64-degree average.

[1] The "fuel sample" in each community was the set of households that completed the community-wide questionnaire and for which reliable date on fuel consumption could be collected for the measurement year. Foley's fuel sample numbered 91 (for both heating and household energy); Munka's fuel sample for household energy was 63; for heating, 41.

Minnesota government and utility companies urged a daytime thermostat setting of 65 degrees. Foleyites set their thermostats to average two degrees above this. The Swedish state recommended daytime temperatures of 68 degrees and nighttime temperatures of 64 degrees. Most Munka Ljungbyans complied with the daytime recommendation but had nighttime temperatures that averaged two degrees above the guideline.

Heating practices should not be discussed for Sweden without reference to patterns of ventilation (*vadring*), practiced daily by most householders. Sixteen of the twenty Munka core households providing information on ventilation aired their houses daily by opening some combination of windows and doors, but only half of them turned thermostats down or off while doing so. Ventilation periods lasted from five to twenty minutes. No relationship was found between duration of ventilation and thermostat adjustment.

Munka Ljungbyans also set their water heaters at higher temperatures. Two thirds (67%) of Munka's core households reported settings of 150 degrees F or higher, while only 15% of Foley's core households did so. In addition, Munka Ljungby water heaters had a greater average capacity than did those in Foley. The mode for Munka's household fuel sample was "greater than 60 gallons but less than 100 gallons. Foley's household fuel sample's mode was 50 gallons, with the 30-gallon size water heater nearly as prevalent.

**Energy-Using Household Activities**

As can be seen from Table 1, Foley and Munka core households demonstrated similar patterns of frequency of use of ovens, ranges, dishwashers, irons, and vacuum cleaners, but differed in other aspects of these activities, and for other activities overall. Differences in average household size should be kept in mind as these patterns are reviewed: Foley's household average was 3.0 persons; Munka's was 2.6 persons.

Many patterns of the Foleyites confirm the stereotype of a more energy-intensive American "lifestyle." While oven and range use are similar in terms of frequency and total energy demand (based on calculations involving temperatures, durations, and types of cooking fuels), ten of Foley's core households contained microwave ovens, while only one of Munka's core did so. Foleyites used their microwave an average of 8.3 times per week. The ovens were used most frequently for tasks that could be performed equally efficiently by the range, such as heating leftovers or water, rather than for the more energy-intensive oven tasks of baking and roast-

**Table 1. Selected activity patterns of core households, Foley and Munka Ljungby.**

| Activity | Community | Average # of times per week | Average duration per time |
|---|---|---|---|
| Oven Use | Foley (N=22) | 3.5 | 90 min. (350 DF mode) |
| | Munka (N=21) | 2.5 | 53 min. (400 DF mode) |
| Range Use | Foley (N=22) | 11.6 | 38 min. |
| | Munka (N=21) | 11.9 | 33 min. |
| Dishwasher Use | Foley (N=10) | 7.6 | not reported |
| | Munka (N=14) | 5.8 | not reported |
| Clothes-Washer Use | Foley (N=22) | 3.6 | 2.4 load per time |
| | Munka (N=20) | 3.7 | 1.2 loads per time |
| Clothes-Dryer Use | Foley (N=21) | 3.8 | not reported |
| | Munka (N=10) | 0.7 | not reported |
| Iron Use | Foley (N=15) | 1.4 | 34 min. |
| | Munka (N=19) | 1.6 | 40 min. |
| Vacuum Cleaner Use | Foley (N=22) | 3.0 | 29 min. |
| | Munka (N=21) | 3.1 | 37 min. |
| Showers | Foley (N=17) | 13.6 | 9 min. |
| | Munka (N=20) | 11.6 | 11 min. |
| Baths | Foley (N=17) | 4.6 | 19 min. |
| | Munka (N=20) | 0.9 | 39 min. |
| TV-Watching | Foley (N=22) | —** | 27 hrs. per week |
| | Munka (N=21) | —** | 13 hrs. per week |
| Radio-Listening | Foley (N=22) | 16.3 | 126 min. |
| | Munka (N=21) | 12.0 | 78 min. |
| Stereo-Listening | Foley (N=21) | 3.5 | 89 min. |
| | Munka (N=19) | 2.2 | 58 min. |

** Calculations of the number of times a TV was turned on each week proved impossible, due to overlapping viewing by household members.

ing, where fuel savings would be realized.

Also, while Foleyites and Munka Ljungbyans used their clothes-washers the same average number of times weekly (3.6 and 3.7 times, respectively), Foleyites averaged 2.4 loads of clothing per laundry episode, while Munka Ljungbyans averaged half that much, 1.2 loads.

Nearly all of Foley's core households, but only half of those in

Munka, owned clothes-dryers. Where choice of dryer versus clothes-line existed, Foleyites chose the dryer 78% of the time. Munka Ljungbyans, in contrast, chose the dryer over the line only 8% of the time.

Foley households recoded a greater weekly average combined number of baths and showers than did those in Munka: 16.0 versus 12.5, respectively. (Six adults in the Munka core household population showered regularly at their workplaces: two adults, 2-3 times per week; two, 4 times per week; and two, 5 times per week.) Where facilities for both showers and baths were available in the home, Foleyites (N=17 households) chose the more energy-intensive bath over the shower more often than did Munka Ljungbyans (N=20 households): 26.5% of the time in Foley versus 7% of the time in Munka.

Finally, Foleyites devoted a much greater percentage of time to TV-watching and to radio or stereo-listening than did Munka Ljungbyans. Foleyites could choose from five different television channels offering nearly round-the-clock broadcasting, while Munka Ljungbyans were limited to three State-owned channels and much shorter broadcasting hours.

Some of the patterns in Munk, however, run counter to the stereotype of careful energy conservation and lesser use of appliances in Swedish households. While Munka Ljungbyans chose showers over baths more frequently than did Foleyites, average durations of both showers and baths in Munka were longer. The average Munka shower lasted 11 minutes, two minutes longer than Foley's average of 9 minutes. Baths in Munka averaged twice as long as those in Foley: 39 minutes, as compared to Foley's 19-minute average.

Munka Ljungbyans ironed and vacuumed the same number of times per week as the Foleyites, but engaged in both activities longer. Although all core households owned irons, a greater proportion of Munka households ironed regularly: 19 of 21, as opposed to 15 of 22. Also, the electric mangle (a device for pressing clothes) is common in Swedish households. Fifteen of the Munka core households owned such a machine; none in the Foley households did.

Two thirds of the core households in each community owned dishwashers, and these appliances were used with equal frequency by both groups. Munka Ljungbyans, however, were more likely to use running water for washing and rinsing the dishes they did by hand. Of all core households, dishwasher owners and non-owners alike, 67% in Munka and 41% in Foley reported using hot running water for washing and/or rinsing dishes.

These findings regarding selection of indoor and water temperatures and household activity patterns are by no means clear-cut. They neither clearly affirm nor deny the stereotypes. Certain patterns are roughly parallel for the two core samples: Foleyites exhibited more energy-demanding behavior in some areas; and Munka Ljungbians did so in others. This mixed pattern implies that structural factors and technology, rather than behavior, predominantly determine Sweden's lesser household energy consumption.

## SOCIAL AND CULTURAL FACTORS THAT PROMOTE ENERGY CONSUMPTION

Core household members were interviewed about the reasons for certain "energy-intensive" choices they made. Examples of these choices are using a clothes-dryer rather than a line, selecting high indoor air or water temperatures, bathing rather than showering, and using hot running water for dishwashing. A wide variety of rationales was given by each group, but within them certain themes could be identified.

### Foley

The energy-intensive choices of Foleyites clustered around three central motivations: attempts to *save time*; needs to *offset the stresses* of daily life; and *compensation for feelings* of powerlessness and anger. These three rationales, furthermore, are culturally legitimated, and individuals citing them are never challenged by peers.

Foleyites try to "maximize" their time, to be as productive and efficient off the job as they are at the workplace. They embody Staffan Linder's (1970) "harried leisure class", attempting to achieve productivity in their "free" time through increasing the numbers of tasks completed, activities engage in, or goods consumed per unit of time. In Foley, being "busy" is a status symbol; the statement "I know how busy you are..." is an ingratiating one. Below, some Foleyites articulate the ways that they perceive time and energy to be related:

- *You could cut energy use tremendously if you had lots of time to spare. But who does? We're always runnin'.*
- *The more energy you use, the less time it takes to*

> *get things done. You gain time through the energy you use.*
>
> — *Using energy saves time. Like taking the car in the mornings, I save 20 minutes. Or her using all the [electric] kitchen stuff. Our whole society is based on that: time. "Let's get moving." Oh boy.*

In order to save time, Foleyites use their clothes-dryers, microwave ovens, other electric appliances, and power mowers; and they drive independently instead of walking, biking, or car-pooling.

In the degree to which Foleyites crowd their daily schedules and experience time pressures is revealed in the comments on the advantages of the dishwasher and clothes-dryer made by two informants:

> — *You can rinse and load the dishes and then be free to go off and do something else. It's really efficient, time-wise. It can wash dishes while you sleep!*
>
> — *With a dryer, you can do things in between loads, rather than commit all that time to the line. You can dry when* you *want to, not just when the weather allows.*

In addition, another informant reported that she was always careful to set the dryer time for more than needed to dry her clothes:

> — *That saves me monkeying around to go down and check to see if they're dry. Otherwise, I used to waste time and waste a trip to the basement!*

A fourth informant said she chose to rinse her dishes with "really *hot*" running water because they "dried faster" that way. This informant did not dry dishes with a towel, but placed them in a draining rack!

Foleyites state that time saved is used "for things we *like* to do" and for participation in the round of activities that community organizations offer. Scheduling interviews with core households was often difficult because of the occupants' unremitting appointments and activities.

This tight scheduling and hectic pace created stress and fatigue. A second rationale for energy-intensive choices in Foley was the need to relax. Baths, "long" showers, TV-watching, and radio and stereo-listening were all cited as ways in which Foleyites could "unwind," "have fun," and protect their "mental health". Foley households contained a greater number of small appliances than did those in Munka, and Foleyites were more likely to express their choices to use these in terms of the enjoyment they offered. One informant said that she bought and used an electric mixer and an electric can opener "because they are *fun*, something you don't get enough of, in the kitchen or anywhere else!"

Many Foleyites felt powerless and angry in the face of rising fuel prices and an uncertain energy future. They trusted neither government nor utility companies, which they felt gave them mixed messages about energy: the denial of an energy crisis by the Reagan administration; conflicting reports on existing fuel supplies and natural resources; and the sharp reversal of former encouragement to consume "penny cheap" fuel. Foleyites felt trapped as dependent fuel consumers. Many chose to reassert a sense of personal power through maintaining or even increasing their household fuel consumption levels. They legitimated these "rebellious" decisions with the third rationale, the argument that they were able to afford the consequences:

— I'm *paying* my *bill. They're not. So I'll do what I damn well please.*

— *I'm not putting a sweater on. I have a* right *to comfort, inside my own house. I pay my own way!*

— *Why should I be cold for two dollars? It's hardly worth it.*

— *Wasting is a luxury. And like any luxury, you* need *it. It's a reward I choose to give myself. My own business.*

Sobel (1981) depicts consumption as a "sacrosanct" area of American life, one in which feelings of power and control are experienced. Foleyites resented attempts to constrain their freedom of consumption, especially since the inalienable right to cheap fuel was being abridged.

**Munka Ljungby**

Time economy figured importantly in Munka also, but not to the extremes that it did in Foley. Munka Ljungbyans were more likely to state that they were *rational* in their use of time, but not that they tried to "save" time. Nor did they engage in the elaborate, down-to-the-minute calculations of the Foleyites. Munka Ljungbyans planned, but Foleyites schemed, to order their daily schedules.

The strongest factors promoting energy consumption among Munka Ljungbyans were *social relations and expectations.* Certain standards of housekeeping, personal appearance, and entertaining are upheld in conformist Swedish society to a degree that amazes American observers. Central to these standards are notions of scrupulous cleanliness and tidiness, and these typified Munka rationales for energy-intensive choices. Munka Ljungbyans used hot water "to get the dishes *really* clean," vacuumed longer to "make *sure* the house is thoroughly clean," and took longer showers and baths to "get all the soap off" and to "get really clean". They used their kitchen fans religiously to get rid of "smells" and "fat" as well as smoke or steam.

More Munka Ljungbyans than Foleyites reported turning up their thermostats when expecting guests. They expressed concern that every guest be comfortable and content, even—or perhaps, especially—the one who "froze" most easily:

> — *It would be* terrible *if* anyone *were cold. Besides, they come in their fine clothes, which aren't at all warm.*

Munka Ljungbyans were more likely to pre-heat their ovens to assure the quality of baked goods, and some Munka informants reported pre-heating an extra twenty to thirty minutes, to be sure that meals for guests would be served punctually.

Compensation for daily stresses and the need to unwind were not the prevalent rationales in Munka that they were in Foley. The Swedish emphasis on a clean and pleasant home environment discussed above stems in part from the need to *counteract an outdoor environment* that is inhospitable for much of the year. In many instances, Munka Ljungbyans chose to consume more energy in order to compensate for long, cold, dark winters. This consumption took the forms of higher indoor and water temperatures, "lots" of lights, "aesthetic" fires in fireplaces, and long, hot baths. Interestingly, several Munka Ljungbyans said they also made many energy-intensive choices in the summer months. They drove their cars more frequently on excursions and used energy to hurry with household tasks so that they could get out into the summer sunshine. Attached to the ceilings of unenclosed porches of some Munka homes were radiant heaters, so that the summertime pleasure of sitting outside in the evening could be extended into the autumn.

As in Foley, the Munka Ljungby themes are legitimate in the eyes of those sharing the culture. No one questions the needs to be rational with use of time; to meet cultural standards of housekeeping, appearance, and entertaining; or to seek redress for an oppressive climate.

## SOCIAL AND CULTURAL FACTORS THAT PROMOTE ENERGY CONSERVATION

### Foley

Certain social and cultural factors are also operant in each community that discourage energy consumption. In Foley, these factors are *budgetary concerns*, *agricultural heritage*, and negative attitudes toward "*waste*."

Foleyites and Munka Ljungbyans shared concerns over household fuel bills, but Foleyites were without the security of subsidies and support offered by the Swedish welfare state. They regarded energy as a household budgetary problem, rather than a broader one; a kind of frontier individualism existed. As with their time calculations, most Foleyites assessed economic costs and tradeoffs involved in their daily choices. Technical and behavioral changes in energy use were made, in the main, to reduce fuel bills.

Foley, like Munka Ljungby, was traditionally an agricultural service center. In many Foleyites a sense of need for conservation exists, but emphasis is on soil and water rather than fuels, including local wood supplies. The farming heritage has a strong residual in many residents, who stressed that they had "always been conservative, of everything!" However, most Foleyites did not have a perspective on larger ecosystems and their places in them.

Interviews with Foleyites revealed another social factor that promoted energy conservation on the parts of many: an abhorrence of "waste." This general distaste usually encompassed waste of household fuels, especially electricity used for lighting. It carried the most gravity with regard to waste of time and food, however.

### Munka Ljungby

In Sweden there was also concern over *fuel bills and budgets* at the household level. Munka Ljungbyans shared a perspective broader than that of Foleyites, however. They expressed concern for the health of their *national economy* and the dangers of its dependence on imported oil. "Solidarity" through efforts to reduce oil consumption was a theme of many interviews.

*Social conformity*, while encouraging energy consumption in order to meet the group standards and expectations discussed earlier, worked in other ways to discourage consumption. Swedes value being "good citizens" and supporting state policies. Munka Ljungbyans were aware of the national committee for energy conservation. They expressed nearly unanimous verbal sympathy with these efforts and guidelines, and some modified their behavior accordingly.

Another cultural factor promoting energy conservation is the *moralistic nature of Swedish society*, characterized as the "society of the 'shoulds'" by one ethnographer (Lofgren 1982, personal communication). More than twice as many Munka core households, fourteen as compared to six in Foley, believed energy use to be a moral issue, and that individuals have a responsibility to the collective good. Swedes believe that there is one "right" position on every issue, no matter what its size, and that once this position has been established, it is the moral duty of the individual to adopt it.

Likewise, the concept of *lagom* pervades Swedish life, buttressing social ethics. Variably translatable as "just right," "in moderation," and "appropriate," *lagom* is applied to the physical characteristics of objects (a jam can be *lagom* sweet: "just sweet enough"), and also to characteristics of social institutions (local government can be *lagom* regulatory) and of individuals (a rude person is sarcastically called *lagom* friendly).

*Lagom* thus connotes both a quantity and a moral judgement on that quantity (Ruth 1984). Individual desires and impulses are brought into conformity with the common spirit, for the common good. Energy and resource use is tempered by the *lagom* code, both in terms of matching energy to task and in delimiting one's "fair share" of the common resources. This restraint prevents any "tragedy of the (Swedish) commons."

A major factor encouraging energy conservation is the Swedes' *love of nature*. Sweden experienced relatively late industrialization at the turn of the century, and the conservative agricultural heritage and ties to the land are stronger there than in the U.S. Personal contact with the natural environment is intrinsic to the Swedish sense of place, time, self, and well-being. Most Munka Ljungbyans expressed concern about the ecological implications of fuel extraction, production, and use. A primary consideration mentioned in discussions of fuel alternatives was which fuel was cleanest and least disruptive to the ecosystem; that is, how *miljo-vanlig* (friendly to the environment) it was. This personal bond with the environment, combined with an ecological awareness, sense of collective identity (at both national and global levels), and assumption of moral responsibility, all work in combination as powerful forces for energy conservation in Sweden.

## ATTITUDES VERSUS BEHAVIOR

More factors promoting conservation, then, are operant in Munka Ljungby than in Foley. Why is it that Munka Ljungbyans express more conserving attitudes but engage in fewer attempts to conserve energy than the Foleyites?

Three major factors in explaining this discrepancy between attitudes and behavior are: economic duress; the role and stature of government; and the self-image of the consumers.

### Economic Duress

Economic considerations were primary in both communities, but critical for Foleyites. Ecological and socio-political factors were expressed concerns of Munka Ljungbyans, but these factors did not induce conservation efforts on their parts. Despite the thread of a general conservation ethic that runs through Foley and characterizes some of its households, the energy conservation efforts of most Foleyites were motivated by economic necessity. They turned down their thermostats, added insulation and weatherstripping to their drafty and often insubstantial housing, put on heavier clothing, monitored appliance and hot water use, and drove less—but grumbled and chafed as they did so.

### Role of Government

Adding time and money-demanding energy conservation practices to the demands already placed on schedules and budgets strained Foleyites even further. In addition, the availability of government fuel assistance was not dependable, and the standards for qualifying for such aid shifted annually. Tax deductions for energy-conserving structural changes in housing were newly available in 1980, and much confusion about them existed. Foleyites did not have access to a central, authoritative source of energy information.

In contrast, Munka Ljungbyans could rely on extremely tight and well-built housing and efficient appliances to relieve them of the need for substantial changes or constant vigilance in their homes. However, a variety of financial aids, both grants and low-interest loans, were available to help cover fuel costs and to make any necessary or desirable structural changes (such as changing furnaces or heating fuel). The Swedish state established special energy advisors to aid consumers. These advisors were placed at the local level of government (the *kommun*, comparable in size to a large U.S. municipality). Most Munka Ljungbyans, therefore, voiced a placid concern about reducing energy consumption, but circumstances did not require them to make changes, either structural or behavioral.

The confidence that most Swedes place in their government and its policies contrasts markedly with the hostility exhibited by Foleyites in the face of energy ambiguities and with their suspicions of collusion between government and "big business" to profit from energy crises. The Swedish welfare state is based upon humanitarian and moral principles, and its policies result from an elaborate process in which academic and scientific specialists, labor unions, businesses, and all public interest groups are consulted. Decisions issued are thus based on consensus, and Swedes can accept them with confidence, knowing that the one "right" conclusion has been reached and that all Swedish citizens will be treated equitably. These policies quickly acquire moral overtones: wasting energy was soon regarded to be "bad" (*daligt*) and "ugly" (*fult*) by the Swedes.

**Self-Image of Consumers**

Foleyites and Munka Ljungbyans had themselves incorporated both the Swedish and American stereotypes described at the beginning of this paper: the temperate Swedes and the gluttonous Americans. Foleyites were apologetic, and the Munka Ljungbyans smug, about their ways of life with regard to energy use. These attitudes helped to precipitate the changes made by Foleyites and to deter changes on the parts of the Munka Ljungbyans. As one Munka informant stated, "Swedes think our way of life is good. We don't have to do anything about it."

Swedes were impressed with what they felt to be the small scale of the size of their country, national population, cars and appliances, and houses and families. Munka Ljungbyans spoke often of the negligible effects of "the tiny demand" and "the small influence" of their households on national energy consumption levels. Such statements were not made by Foley informants, although they did compare their consumption with that of other types of household consumers; namely, city-dwellers and the wealthy.

The desire to reduce energy consumption did not motivate Munka Ljungbyans to choose the clothes-line over the clothes-dryer or bicycle or walk instead of drive. They chose the clothes-line, for the most part, because they did not have a dryer, or because drying clothes outside conforms to the cultural emphasis on ventilation of the household and its contents, or because they were not satisfied with the stiff clothes produced by the most commonly-owned type of clothes-dryer in Sweden, the *torkskap* or "drying cupboard," in which clothes are draped over rods and blown dry. Bicycling or walking was chosen over driving mainly because it is "beautiful" (*skont*) to exercise, rather than in order to reduce gasoline consumption.

The factors promoting energy conservation in Sweden—ecological awareness and concern, love of nature, strong national and global identification, moralistic outlook, and the *lagom* code—shape more directly the decisions of Swedish policy makers than those of individual consumers. The latter can depend passively on their energy-efficient technology and their government's social welfare and energy policies to control energy consumption and to maintain an equitable security.

Symbols of the energy contexts in Foley and Munka Ljungby might be the American "crazy quilt" and the "Swedish modern" chair. The quilt is home-made, and richly textured, but unpredictable, without an overall design or plan, and a little too short: Foleyites struggle to keep it tucked in around their feet! The Swedish chair, in contrast, is streamlined, beautiful, and unambiguous, in which, due to skilled designers and engineers, Swedes relax with confidence.

**References**

Darley, J. M., C. Seligman, and L. J. Becker. "The Lesson of Twin Rivers: Feedback Works." *Psychology Today* April 1979.

Doernberg, A. "A Comparative Analysis of Energy Use in Sweden and the U.S." Brookhaven National Laboratory, Report GNL-20439. Upton, NY, 1975.

Gaunt, L. and A. Berggren. "Household Habits and Energy Consumption." The National Swedish Institute for Building Research, Gavle, Sweden, 1983.

Goldstein, R., et al. "Analysis of Residential Energy Use." Heat Transfer Laboratory Report #114. Institute of Technology, University of Minnesota., Minneapolis, Minnesota, 1979.

Hambraeus, G. and S. Stillesjo. "Perspectives on Energy in Sweden." *Annual Review of Energy.* 1977.

Linder, S. *The Harried Leisure Class.* New York: Columbia University Press, 1970.

Lofgren, O.. Department of European Ethnology. University of Lund, Sweden. Personal communication, 1982.

Lundstrom, E. *Energiforbrukning i Smahus.* Institutionen for Byggnadsekonomi och Byggnadsorganisation. Stockholm: Royal Technological Institute, 1980.

Ruth, A. "The Second New Nation: The Mythology of Modern Sweden." *Daedalus,* Spring 1984.

Schipper, L. and A. J. Lichtenberg. "Efficient Energy Use and Well-Being: The Swedish Example." *Science,* 194, 1976.

Sobel, M. *Lifestyle and Social Structure.* New York: Academic Press, 1981.

# Residential Hot Water: A Behaviorally-Driven System

*Willett Kempton*
*Princeton University*

## INTRODUCTION

As improved space conditioning and building shell technology make possible dramatically reduced energy requirements, hot water grows in relative importance for residential energy consumption. Most previous studies have viewed the water heater as an energy-consuming appliance. This chapter views the water heater as an intermediate conversion device, and views the hot water taps as the energy-consuming appliances. Apart from standby losses, hot water energy consumption is behaviorally driven by water-use events ranging from a few seconds to over 20 minutes in duration, and from less than 1/10 liter to 100 liters in volume. In order to understand hot water consumption, one must examine these water-use events, the needs they serve, and the efficiency of the system in meeting those needs. This chapter represents a first descriptive step toward those ends.

## MEASUREMENT TECHNIQUES

Our water measurements are part of a comprehensive instrumented monitoring project measuring many energy variables in single-family residences (Weihl, Kempton, and DuPage 1984). The instrument data are supplemented by intensive open-ended inter-

*Energy Efficiency: Perspectives on Individual Behavior*

views with the study families. The cost and intrusiveness of the data collection has precluded both large samples and random selection. This chapter draws from data on seven houses, instrumented for 7 to 18 months. All houses had storage water heater tanks, as is typical in the U.S.; two were heated via electric resistance and the remainder by gas. The variables of interest are the volume of water used, temperature of water entering and leaving the water heater, the amount of gas or electricity used by the water heater, and the temperature of the hot water pipe at each tap using water. These data are recorded by a programmable microprocessor-controlled field instrument with one-minute resolution. Our project's choice of measurements was a compromise between what we wanted to know and what was practical. We would have liked to know the volume used at each tap, but expense and installation difficulties precluded installing a volume meter in every pipe. In this paper I use a method, described elsewhere (Weihl and Kempton 1985), by which a single volume meter and temperature rises at the taps allow inference of which tap used how much water.

## MEASURING WATER EVENTS

Energy consumed by storage water heaters is lost in conversion to heat and in losses through the tank walls and adjacent pipes ("standby loss"). The remainder is delivered through the piping to the tap. Conversion and standby losses have been extensively analyzed in energy conservation research on water heaters (Hirst and Hoskins 1977; Wilson 1978; Palla 1979). While advances in our understanding of heat losses have resulted in improved water heaters, the amount of energy lost is usually smaller than that embodied in the hot water drawn from the tank.[1] Thus, in order to effectively address the largest component of hot water heating energy, one must learn more about the residents' consumption events rather than the water heater itself.

For a behavioral analysis of hot water, I consider the primary unit as not a joule of fuel or a liter of water, but as a "water event." I define a water event as a contiguous flow of hot water bounded by at least one minute without hot water use before and after. At least one minute is required between events due to technical constraints on our recording devices. This definition success-

[1] A spectacular exception to the typical single-family case is seen in a recent study of a multi-family building with an inefficient boiler and a long recirculating distribution loop. Delivered hot water was only 12% of fuel consumed, and only 27% after replacement of the boiler (DeCicco and Dutt 1986).

fully isolates behaviorally coherent units, such as showering, washing hands, and filling a sink with water, as appropriately-delimited "events." It is not as appropriate for some events, which may overlap or be artificially split. For example, the cycles of a dishwasher will be artificially split into distinct events. Conversely, a sink may be used twice in succession for two different purposes. If the uses occur within a minute of each other, they will be inappropriately combined into a single event. Nevertheless, for most analyses one minute resolution has proven satisfactory.

Temperature rises at individual taps allow attribution of events to specific taps, and thus to approximate categories of usage. Since restricted access to bathrooms in three houses necessitated a single probe for bathroom sink and tub, bathing and sink use are distinguished using a 15 liter cutoff.[2] The measurement equipment is described in more detail elsewhere (Weihl et al. 1984; Weihl and Kempton 1985); suffice it to say here that we think this method offers a reasonably accurate measure of hot water behavior at moderate cost, using available equipment.

The energy content of the water transmitted from the tank to the plumbing system is computed from the volume and temperature of the water leaving the tank,[3] based on a 12°C inlet water temperature.[4] Heat arriving at the tap is computed similarly. While most

---

[2] Bathing is considered to be 15 liters and above; bathroom sink use is considered to be less than 15 liters. This cutoff criterion is based on a distributional analysis of bathroom water use in the four houses in which we monitored the two bathroom uses separately. Tub and shower use peaks around either 40 liters or 60-70 liters, depending on the house, while bathroom sink use peaks at 1 liter and below. At a 15 liter cutoff, showers are incorrectly attributed to the sink only for a few of the quickest showers (seen only in house 2), and sink use is misattributed to the shower only for rare long uses such as prolonged hair washing.

[3] Given continuously measured $V_m$ (volume in liters at minute m) and $T_m$ (tank water temperature at minute m), an assumed constant $T_i$ of 12°C (inlet water temperature), and a conversion factor of 4.14 kJ/l°C, heat embodied in the water leaving the tank is computed as:

$$4.14 \sum_{m} (T_m - T_i) V_m$$

[4] We ignore seasonal variation in inlet temperature. Inlet temperature was measured only at house 5, using a daily minimum measured about 8 meters from the house inlet. The year-round range of monthly average inlet temperatures was from 10.3°C to 15.4°C. At a typical hot water temperature of 58°C, assumption of constant inlet temperature would cause an error of less than 8%. Some municipal water systems exhibit much more variation in inlet temperature (Hirst, Goeltz, and Hubbard 1986); analysis of such systems could not assume a constant inlet temperature.

**Table 1. Hot Water Characteristics of Study Houses**

| House | 0 | 1 | 2 | 3 | 4 | 5 | 7 |
|---|---|---|---|---|---|---|---|
| Residents | 2 | 5 | 2 | 3 | 4 | 6 | 4 |
| Sampled days | 89.2 | 50.9 | 183.1 | 123.0 | 96.8 | 62.5 | 21.4 |
| Sampled events | 1268 | 2076 | 2356 | 2151 | 2777 | 3042 | 621 |
| Events/day | 14 | 41 | 13 | 17 | 29 | 49 | 29 |
| Liters/day | 130 | 199 | 89 | 266 | 315 | 290 | 505 |
| Liters/day/person | 64.8 | 49.8 | 44.5 | 88.7 | 78.8 | 48.2 | 126.4 |
| Maximum tank temperature °C* | 53° | 65° | 59° | 62° | 52° | 67° | 46° |
| Tank water MJ/day** | 15.8 | 35.9 | 13.0 | 41.2 | 44.3 | 52.1 | 67.4 |
| MJ/day/person | 7.9 | 7.2 | 6.5 | 13.7 | 11.1 | 8.7 | 16.9 |

* Maximum temperature measured during sample period (°C).

** Energy embodied in water leaving tank.

of the measurements are very accurate, the heat calculations are imperfect due to our pragmatic choice of temperature measurement at the pipe surface. Thus, the heat data should be treated as an approximation.

## DAILY WATER USE

Table 1 lists the data used for the analysis of this paper. For each of the houses included in the study, I have tried to include at least 2000 water events, and at least 50 contiguous days of data. Some compromises in these criteria had to be made due to equipment failures and limits on operating time in each house. (House 6 is excluded due to multiple equipment problems.) Table 1 shows that Houses 1 and 5 use hot water most frequently, while House 7 uses the most hot water. House 7's usage is largest whether computed by liters or megajoules, and whether computed by household daily total or per person. More generally, Table 1 shows considerable variation in hot water use among houses—variation which is only partly due to the number of residents. Some idea of the cultural and family differences responsible for this variation will be discussed below; further insights can be gleaned from Weihl's analysis of daily family schedules in these houses (Weihl 1986).

**Table 2. Disaggregation of Daily Hot Water Use by Location: Events/Day and Liters/Day**

| House | 0 | 1 | 2 | 3 | 4 | 5 | 7 | Mean |
|---|---|---|---|---|---|---|---|---|
| Household Total* | | | | | | | | |
| Events | 14 | 41 | 13 | 17 | 29 | 49 | 29 | 27 |
| Liters | 130 | 199 | 89 | 266 | 315 | 290 | 505 | 256 |
| Kitchen | | | | | | | | |
| Events | 7.6 | 20.0 | 7.4 | 8.8 | 17.0 | 30.1 | 15.0 | 15.1 |
| Liters | 37.9 | 67.2 | 16.6 | 55.0 | 104.8 | 96.2 | 102.9 | 68.7 |
| Bathroom bathing | | | | | | | | |
| Events | 1.1 | 1.6 | 1.4 | 2.4 | 2.3 | 0.8 | 3.6 | 1.9 |
| Liters | 70.8 | 68.6 | 58.2 | 169.1 | 91.3 | 47.4 | 226.5 | 104.6 |
| Laundry | | | | | | | | |
| Events | 0.7 | 2.9 | 0.6 | 1.2 | 2.9 | 5.1 | 3.4 | 2.4 |
| Liters | 6.2 | 33.8 | 9.2 | 30.1 | 87.7 | 115.5 | 150.2 | 61.8 |
| Bathroom sinks | | | | | | | | |
| Events | 3.4 | 15.1 | 3.1 | 4.3 | 6.0 | 12.1 | 6.3 | 7.2 |
| Liters | 13.3 | 28.2 | 4.3 | 11.0 | 29.1 | 30.0 | 25.6 | 20.2 |

* Liters at individual locations do not quite sum to the household daily total because the locations of a small percentage of events could not be identified.

Total daily water consumption is disaggregated by point of use in Table 2, which compares houses by number of events and by liters. In every house, the kitchen sink accounts for the greatest number of events. In one house it accounts for the greatest volume as well, though in the majority bathing accounts for the greatest volume, with kitchen second. As with the total daily household use, we find large variation among houses in volume and events for each use. Bathing varies by a factor of five-to-one. The variation is not reduced (it actually increases to seven-to-one) when calculating volume per person, indicating that inter-house differences are due to personal and cultural factors as much as to family size. Kitchen and laundry uses show a greater than six-to-one variation in volume, not significantly reduced by calculating per person.[5] That is, the variation across houses in individual water use behaviors is even greater than the variation across houses in total hot water use.

[5] Event-disaggregated MJ or MJ/person can be calculated from liters in Table 2, by constructing house-specific approximate conversion factors using Table 1. The °C in Table 1 should not be used, since it is a maximum, and the MJ figures are based on continuous temperature measurements.

## BATHING

Consistent with previous studies such as Wilson (1972), Table 2 shows that bathing averages about 40% of residential hot water; in most houses it is the largest single use. Previous studies have not reported the large variance among individual houses. If some households average 40 liter showers, why do others require 70 liter showers? The amount of energy involved in bathing makes further investigation worthwhile. I will analyze two individual cases to illustrate some of the cultural and psychological determinants of hot water use. But first, a caveat is in order.

Frequency of bathing, and some aspects of volume and duration of bathing, are issues of personal preference and culture, and are surely not issues that energy conservation programs should try to influence. Nevertheless, there are good reasons for studying bathing water use. The characteristics of personal bathing may be helpful to know in calculating the effects of technical efficiency improvements, such as low-flow shower heads, or in predicting the social barriers to adoption of such technologies. Further, if information were publicly available about the range of possible bathing behavior and the range of consequences for energy cost, any individuals who wished to let economics affect their own decision-making would be free to do so.

By dividing liters/day by events/day, one finds that the houses in our sample divide into two groups: three houses have about 40 liter bathing, and four houses have 60 to 70 liter bathing. To illustrate two clearly contrasting cases, the following figures will compare the lowest-volume bathing, in House 2 (average 41 liters/event) with the highest-volume bathing, in House 3 (average 70 liters/event).

House 2 is occupied by a couple near retirement, who have a regular daily schedule. In the interview, Bill reports taking a "two to three minute" shower at 6:30 each morning, and Renee reports a "five minute" shower at 7:00 or later. The first eight showers of January 1984 are shown in Figure 1, which plots the liters flowing at each minute against the number of minutes since the beginning of the shower. Solid lines indicate the first shower of each day and dotted lines indicate the second. The pattern is entirely consistent with their verbal reports, although each of them underestimates duration slightly. Bill's showers are highly consistent from day to day, while Renee's are more variable. (The shower that extends off the right edge of Figure 1 occurred the morning after New Year's

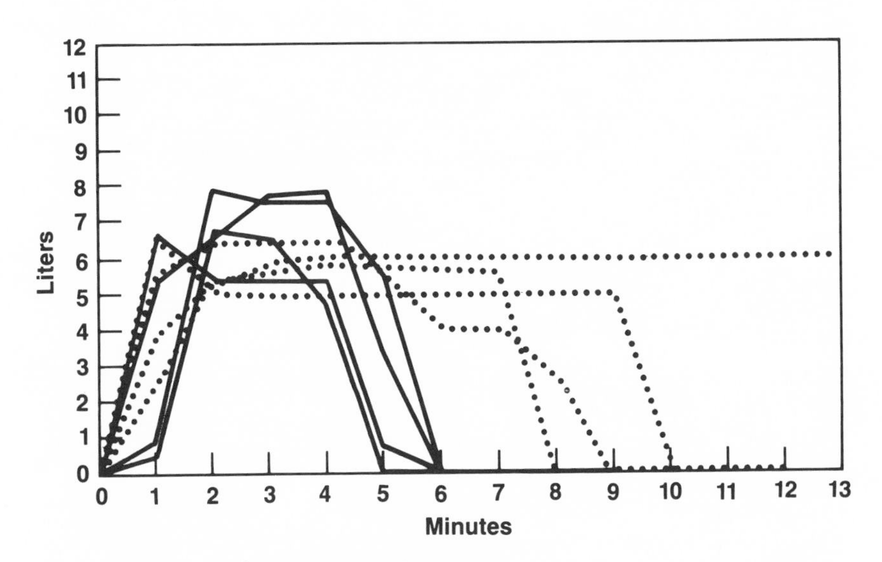


**Fig. 1. Showers of Bill and Renee in house 2.**

Eve.) Bill's showers average about 20 liters of hot water, or about 3MJ energy content.

The figure shows clear individual differences, and in this household the residents are aware of the differences. The following discussion, which occurred spontaneously in response to our interview questions about the time of day showers are taken, raises questions of waste, thrift, enjoyment, and cleanliness, and illustrates the strains that can arise when one household member tries to limit the energy use of others.

*Renee:* I usually take 'em in the morning, but it's not until about 9:00.

*Bill:* You'd be interested to see how long she takes showers. (laugh)

*Renee:* I keep telling you it's longer than you take a shower (laugh) (pause) I mean, I've got a lot more hair to wash than you have.

*Bill:* Right. And it's all kinds of other things, and I'm (pause) almost ready to accuse her of just standing under this hot water and just ... you know ...

*Renee:* Oh, it feels good.

*Bill:* Right.

*Interviewer:* I think I do that occasionally myself.

*Bill:* I do it if I'm at a commercial place, you know, like [name] Community College. When somebody else is paying. I never ...
*Renee:* You don't pay that much for our energy.

For Bill, quick efficient showers fit in with a general attitude toward frugality. For example, this household also follows a careful and efficient heating thermostat setback schedule. Bill seems to value efficiency for its own aesthetics, and the family also has some financial motivation–they live comfortably, but say that they do not have extra money at the end of each month. When challenged (above) Renee is a little defensive about her showers; she makes a joke implying that Bill does not clean adequately. However, one has the impression that she will continue doing what she wants. For her, the shower is both functional and something to enjoy.

Bill's quick showers also fit his brisk morning routine. He reports getting up at 6:30, "then I take a shower and start breakfast," being ready to eat "about seven" and "I walk out the door ten minutes after seven." Together, the above interview data make sense of House 2's place at the low end of the bathing water use distribution.

I next examine the case of House 3, which had the highest-volume bathing. In our interview at House 3, Greg described his 5:20 am showers as lasting "five to seven minutes," while his wife reports hers as being later and lasting only 5 minutes. The instrument data show consistent 9 to 12 minute showers occurring around 5 am, with shorter showers occurring later in the morning. Figure 2 superimposes four of the long showers from House 3, which shows a consistent pattern generated by a single individual, presumably Greg. These 9 to 12 minute, 90 to 100 liter showers contrast sharply with Bill's showers (seen in the solid lines in Figure 1). House 3's bathing water use is further increased because Greg sometimes takes a second shower after returning from work.

Greg's showers are taken in an upstairs bathroom, which is a 13 meter pipe run from the water heater. Most readers of this chapter will recognize Greg's pattern of hot water adjustments in Figure 2: high initial flow to warm the water up quickly, followed by adjustment down to a comfortable temperature, followed later by a gradual return to higher flow as the tank's hot water is diluted by incoming cold water. These 100 liter showers would be expected to tax House 3's 127 liter (30 gallon) water heater. This graph suggests that Greg may have limited the length of his habitual shower in order to avoid emptying the tank. Judging from the sudden

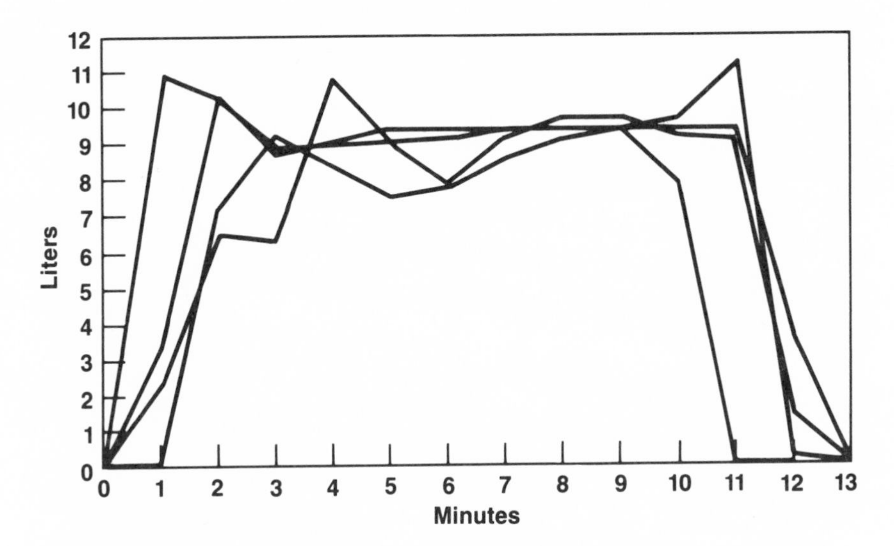


**Fig. 2. Greg's showers at house 3.**

spike toward more hot on the upper right of one shower, we inferred that he sometimes has to finish promptly to avoid a chilly rinse. In a later interview, he verified our inference.

Figures 1 and 2 illustrate different people using widely varying resource quantities to perform the same function. Actually, one could argue that the function is not the same. Bill is a white collar government employee, while Greg works in an automobile factory. In the interview, Greg reports taking his 5:20 am shower as part of "trying to wake up" because "that is too early." Greg noted that their summer gas bill seemed high, and that he turned the water heater temperature down to try to save money. However, cost does not seem to be a major concern at House 3—in responding to our thermostat questions, both Greg and his wife say they do not believe in being uncomfortable to save money. From these case study data, one would conclude that Greg is unlikely to be responsive to appeals to conserve hot water via shorter showers as he groggily adjusts the taps in the early morning.

What motivation does our economic system inject into this situation? Houses 2 and 3 both have gas-fueled water heaters. From our heat measurements, liters can be converted to MJ heat contained in the water leaving the tank. Bill takes 3 MJ showers and Greg takes 15 MJ showers. Based on interview data, I estimate 365 showers/year for Bill and 440 for Greg. Assuming 60% gas water heater efficiency and current rates of $0.50/ccf, Bill pays $9 annually for his showers, while Greg pays $54. Based on this household's other energy decisions, we felt that Greg would be willing to pay this cost, were he presented with the information (he later verified this). For comparison, we can assume a more costly environment: electric water heating at the high marginal rate of $0.15/kWh (and assuming 100% conversion efficiency). In this environment, Bill would pay $44 annually for showers and Greg would pay $277; this expense might interest Greg more in shortening his showers or installing a more efficient shower head.

In our second interview, we presented Greg with Figures 1 and 2, and the present gas costs and projected electric costs. He felt that his present costs were well worth it, but said he would consider changing his use of the shower if the cost were a couple of hundred dollars. However, economic motivation requires that the consumer would know what proportion of his utility bill were going to showers, a dubious assumption (see Kempton and Montgomery 1982). American consumers more commonly underestimate the amount of energy required for hot water, though they may overestimate it in countries with point-of-use hot water heaters (Kempton et al. 1985). We showed all occupants our cost figures at the end of the study. Consistent with those earlier findings, the occupants were particularly surprised by the high cost of electric resistance water heating. In short, the potential for economic motivation exists, depending on fuel type and marginal utility rates, but information failures could prevent that potential from being realized.

## HOW MUCH TANK HEAT IS DELIVERED TO THE WATER TAPS?

The efficiency with which hot water is distributed within a house is determined by an interaction between water event behavior and the physical distribution system. This section compares heat arriving at the water taps with heat sent from the water tank. Both measurements depend upon temperatures measured at the pipe surface, and thus they only approximate the temperature of the water within the pipe. Heat contained in tank water is lost through conduction to the pipe, and then by conduction and radiation to the environment. Heat is lost especially on short water events, when hot water leaves the tank but does not reach the tap, then cools to room temperature before being used. To measure such losses, I define "hot water distribution efficiency" as the heat

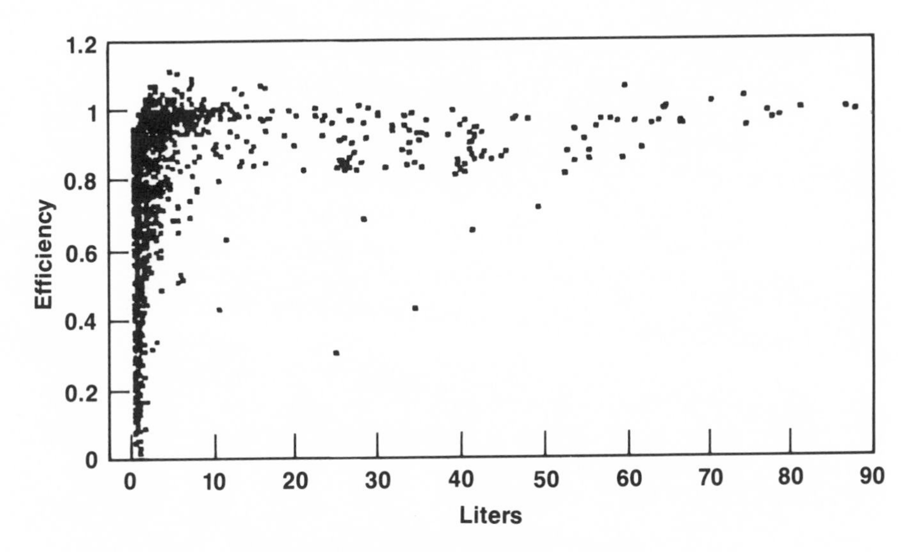


**Fig. 3. Hot water distribution efficiency at house 5.**

delivered to the tap, divided by the heat leaving the tank.[6]

Distribution efficiency could only be accurately calculated for Houses 5 and 7, where temperature probes were mounted close to every tap in the house. The following analysis used House 5, for which most data exist. House 5 is a wood-frame two-story dwelling, built about 1970. It has the water heater and laundry in the basement, a kitchen and half bath on the first floor, and a full bath on the second floor. Pipe runs are 1/2" copper pipe and entirely inside the heated space.

Figure 3 plots volume against efficiency for 1392 water events during the first 26 days of our sample at House 5. If more than 10 liters of hot water were used, Figure 3 shows that usually over 80% of its heat would arrive at the tap. For events under ten liters, there is considerable variation in efficiency. Figure 4 expands the 0 to 10 liter portion of the previous figure, showing more clearly that while low-volume events vary greatly in efficiency, efficiency improves markedly above a liter or two. Other than volume, the two factors that cause this variability are the proximity of the tap and the time since that tap was last used. If water is drawn twice in succession, the second event will be more efficient.

[6] Single-event efficiency can be above 1.0, as seen in Figures 3 and 4, due both to transients and to hot and cold water mixing in the storage tank.

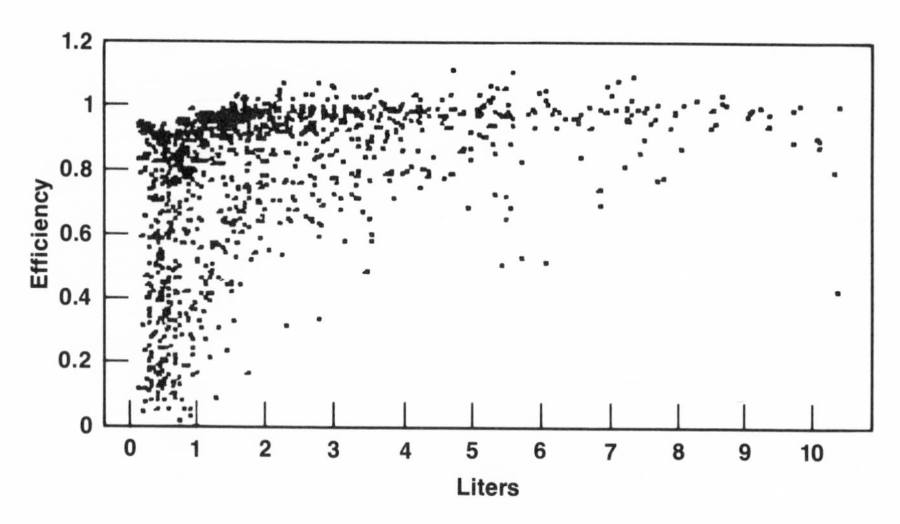


**Fig. 4. Hot water distribution efficiency (expanded scale).**

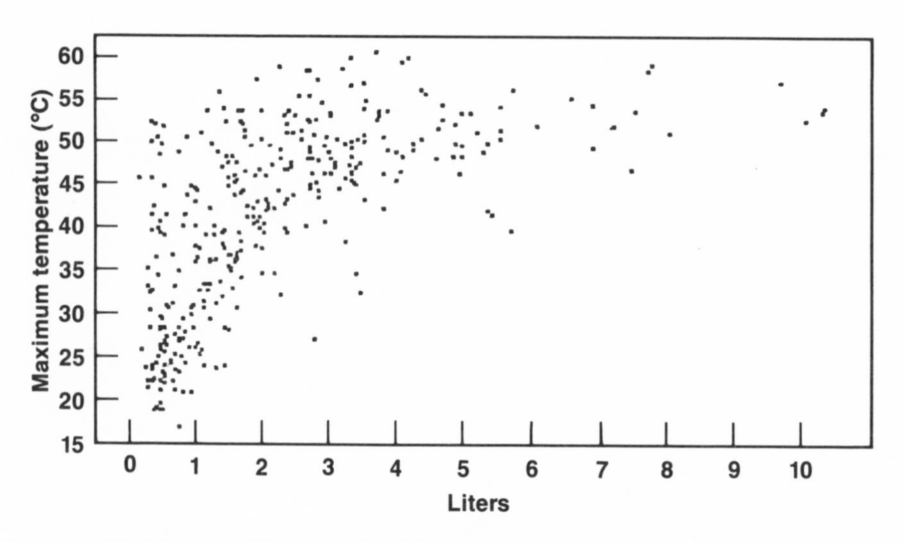


**Fig. 5. Maximum temperature delivered to bathroom sinks at house 5.**

The short events primarily comprise water use at the kitchen and at the bathroom sinks. In this house, as in many American houses, the kitchen sink is much closer to the water heater than are the bathroom sinks. Consequently, small-volume uses are less efficient at the bathroom sinks than in the kitchen.

For the 364 events occurring at bathroom sinks during this period, Figure 5 plots the maximum temperature ever reached by the water. That is, if three liters were drawn, and the water reached 50°C only after the third liter, then 50°C is plotted for that event. Even using this maximum measure, Figure 5 shows that low-volume events usually do not achieve very high water temperatures. In fact, events under two liters most frequently do not even reach body temperature (37°C).

In these low-volume events, the occupant draws hot water from the tank, but hot water is not delivered to the tap. Why is the resident using the hot tap in these instances? It may just be an unconscious habit of reaching for the hot tap. It may be a desire for heat, coupled with an unwillingness to wait. Or it may be insurance that the water will never be *below* room temperature. At low volumes, the cold tap, like the hot tap, will often deliver room-temperature water. However, the cold tap—unlike the hot tap—may deliver uncomfortably cold water after a large water event elsewhere.

Having demonstrated that people's behavior is inefficient for small water uses, the next question is, "Does it matter?" Table 3 gives the efficiencies for water events at each location for the two houses with usable heat data. As expected, the bathroom sinks show the lowest efficiencies. The right column under each house gives the proportion of the water volume that is consumed at that tap. Thus, for instance, although bathroom sinks are inefficient, they consume little water; therefore they have a lesser effect on

**Table 3. Distribution Efficiency, Weighted by Volume**

| | House 5 | | House 7 | |
|---|---|---|---|---|
| | Efficiency | Volume (%) | Efficiency | Volume (%) |
| Kitchen | 0.97 | 33 | 0.84 | 20 |
| Bathroom sinks | 0.75 | 10 | 0.76 | 5 |
| Bathroom tub and shower | 0.95 | 16 | 0.96 | 45 |
| Laundry | 0.90 | 40 | 0.83 | 30 |
| Weighted efficiency | 0.91 | | 0.88 | |

overall efficiency. When efficiency is weighted in this way, it is surprisingly good, about 90% in each case in Table 3. Note that this figure refers only to the house piping system. Overall efficiency is much lower when including losses at the water heater and in utility transmission and generation.

The conclusion of these efficiency calculations is that although efficiency is low for a large number of events, it is high in proportion to total water volume. Thus, although the inefficiency of sinks is inconvenient—residents may frequently get cold water when they want warm water—it is not expensive, since short events comprise a small proportion of total volume. Close proximity of the kitchen sink to the water heater is a major help, since the kitchen consumes much water in the form of many small events. (Presumably builders long ago discovered that home buyers dislike waiting for warm water at the kitchen sink—in this case optimizing the design for convenience would have promoted efficiency.)

Although efficiency of the distribution system looks surprisingly good, one should remember that the two houses tested were fairly new, with copper plumbing making direct runs in heated interior space. Efficiencies would be lower in other houses with less favorable conditions.

## CONCLUSIONS

This chapter examined behavioral aspects of hot water consumption using high-time-resolution instruments and open-ended interviews. In single-family residences, water heating energy consumption is determined primarily by behavior. Thus, the hot-water taps were viewed as the energy-consuming appliances and the water heater as a conversion device. Energy consumption is seen as a series of water event behaviors rather than as tank burner time. Hot water consumption varies greatly among residences, in total household use and even more in consumption for each type of water event.

While this research is only a first exploration of hot water behavior, some policy implications can be brought out. The large variation across individuals in each type of hot water use reveals a personal, cultural, or discretionary component even larger than the variation in total household consumption would suggest. Bathing is the largest single hot water end-use within the residence. Using in-depth case studies, I examined the low usage of a person who valued frugality, efficiency, and a brisk morning schedule, and compared it with the five times greater usage of a person with a physically demanding job who used showers not just for bathing but also "to wake up". While the small number of houses studied here cannot provide quantitative indicators, these case studies suggest which social and personal variables might be investigated to find quantitative indicators of hot water consumption.

Although the potential energy savings of behavioral changes in hot water use appear large, conservation programs addressing these components may be more controversial than technical fixes such as new equipment or retrofits. Some social-psychological techniques, such as having one person model efficient showering behavior, have been shown effective in public facilities such as gymnasium showers (Aronson and O'Leary 1983). It is not clear that these techniques could be applied to a program for single-family residences. At a minimum, programs should make it easier for people to learn the cost of hot water (or better, the cost of each hot water end-use), and the potential savings of the alternatives. While information alone often has little effect (Stern and Aronson 1984), consumer underestimates of hot water cost, especially for electric resistance water heaters, suggests that we do not currently even enjoy the conservation effects that market forces would provide.

**Acknowledgements**

*This research was conducted as part of the author's work with the Family Energy Project at Michigan State University. It was supported by grants from the National Science Foundation (BNS 82-10088), the Michigan State University Agricultural Experiment Station (Project 3152), and the Kellogg Biological Station Small Farms Project. Writeup and analysis were also supported through the New Jersey Energy Conservation Laboratory at the Center for Energy and Environmental Studies, Princeton University. I am grateful to Joanne Keith and Jeffrey Weihl for collaboration in data collection, and to John DeCicco, Eric Hirst, and Jeffrey Weihl for comments on this chapter.*

**References**

Aronson, E. and M. O'Leary. "The Relative Effectiveness of Models and Prompts on Energy Conservation: A Field Experiment in a Shower Room." *Journal of Environmental Systems* 12:219-224, 1983.

DeCicco, J. and G. Dutt. "Domestic Hot Water Service in Lumley Homes: A Comparison of Energy Audit Diagnosis with Instrumented Analysis." In *Proceedings from the ACEEE 1986 Summer Study on Energy Efficiency in Buildings.* Washington, D.C.: American Council for an Energy-Efficient Economy, 1986.

Hirst, E., R. Goeltz, and M. Hubbard. "Determinants of Electricity Use for Residential Water Heating: The Hood River Conservation Project." Manuscript, June 1986.

Hirst, E. and R.A. Hoskins. "Residential Water Heaters: Energy and Cost Analysis." *Energy and Buildings* 1:393-400, 1977.

Hopp, W.J. and W.P. Darby. "Household Water Conservation: The Role of Indirect Energy Savings." *Energy* 5: 1183-1192, 1980.

Kempton, W., C.K. Harris, J.G. Keith, and J.S. Weihl. "Do Consumers Know 'What Works' in Energy Conservation?" *Marriage and Family Review* 9:115-133, 1985.

Kempton, W. and L. Montgomery. "Folk Quantification of Energy." *Energy* 7: 817-827, 1982.

Palla, R.L., Jr. *The Potential for Energy Savings with Water Conservation Devices.* NBSIR 79-1770, U.S. Dept. of Commerce, National Bureau of Standards, Washington, DC, 1979.

Stern, P.C. and E. Aronson, Eds. *Energy Use: The Human Dimension.* New York: W.H. Freeman and Company, 1984.

Weihl, J.S. "Family Schedules and Energy Consumption Behaviors." In this volume.

Weihl, J.S. and W. Kempton. "Residential Hot Water Energy Analysis: Instruments and Algorithms." *Energy and Buildings* 8: 197-204, 1985.

Weihl, J.S., W. Kempton and D. DuPage. "An Instrument Package for Measuring Household Energy Management." In B. Morrison and W. Kempton, Eds., *Families and Energy.* East Lansing: Michigan State University, Institute for Family and Child Study, 1984.

Wilson, R.P. "Energy Conservation Options for Residential Water Heaters." *Energy* 3:149-172, 1978.

# Thermostat Management: Intensive Interviewing Used to Interpret Instrumentation Data

*Willett Kempton*
*Princeton University*
*Shirlee Krabacher*
*Michigan State University*

## INTRODUCTION

Thermostat setting was a major component of the public energy conservation appeals by U.S. and state governments in the 1970s. It is also mentioned prominently by consumers when they discuss energy conservation.[1] In most surveys, consumers report that they have lowered their thermostat settings since the 1973 oil embargo (Vine 1984; and chapter by Vine in this volume). The actual energy savings of these reductions has been estimated by Fels and Goldberg (1984). Using indirect but reasonable methods they estimate that in New Jersey residences, fully 50% of the decrease in natural gas consumption between 1972 and 1980 can be attributed to lower thermostat settings. From the Fels and Goldberg data,



---

[1] Evidence that consumers consider the thermostat important derives from open-ended questions in which they mention it as one of the first and most frequently-given energy conservation strategies (Kempton et al. 1984). This seems to be part of a general pattern in which consumers overestimate the effectiveness of management strategies and underestimate the payback of capital investments (Kempton et al. 1984).

Kempton estimates that American households are currently saving $5 billion annually due to changes in thermostat management since 1973 (1986). In short, consumers choose thermostat management as a primary energy conservation strategy, and this strategy is proving effective.

To understand household thermostat setting we must recognize that it is a conscious and deliberate behavior resulting from complex household decisions. It may involve a set of rules concerning the proper setting for particular activities and times of the day. Even a household that employs a single thermostat setting throughout the heating season—seemingly a lack of management—must decide which setting is "correct," when in the fall to begin active use of the thermostat, and when in the spring to set it back to the lowest point again. The selection of the thermostat setpoint is the result of a potentially intricate decision-making process in which comfort and attitudes concerning the necessity of heat are weighed against economic constraints and, sometimes, morality or patriotism. Many families consciously develop policies for setting the thermostat. Some make specific individuals responsible for setting the thermostat or for checking it periodically to make sure it is set properly.

## METHODS

Researchers have found the study of residential thermostat behavior both interesting and frustrating because it is notoriously difficult to measure. Most studies have relied on self-reports, lower settings.

In order to better understand thermostat management we selected a small sample of diverse households for intensive study. We installed an electronic recorder in each house, which continuously monitored the thermostat setting, exterior and interior temperatures, furnace run-time, window and door openings, and other energy variables over a two year period (Weihl et al. 1984). After the instrument had been recording for several months, we conducted an interview to determine family schedules, thermostat setting policies, and reasons for management strategies. Additionally we maintained extensive contact with the families during installation of the instruments and during periodic visits for data tape changes and equipment servicing. Thus, we have comprehensive data from diverse sources, but we have sacrificed the advantages of a large sample and systematic sampling techniques. Our sample of

**Table 1. Mean Thermostat Settings by House and by Season***

| | House | | | | | | |
|---|---|---|---|---|---|---|---|
| | 1 | 2 | 3 | 4 | 5 | 6 | 7 |
| Fall °F | 71.3 | | 65.1 | 60.7 | | | |
| [°C] | [21.8] | NA | [18.4] | [15.9] | NA | NA | NA |
| (SD) | (3.4) | | (5.2) | (5.1) | | | |
| Winter °F | 73.6 | 66.5 | 69.2 | 63.6 | 73.5 | 69.5 | 69.3 |
| [°C] | [23.1] | [19.2] | [20.7] | [17.6] | [23.1] | [20.8] | [20.7] |
| (SD) | (1.9) | (3.9) | (1.1) | (4.4) | (5.5) | (2.4) | (5.6) |
| Spring °F | 60.2 | 66.9** | 64.2 | 59.2 | 60.4 | 65.0 | 68.4 |
| [°C] | [15.7] | [19.4] | [17.9] | [15.1] | [15.8] | [18.3] | [20.2] |
| (SD) | (12.0) | (3.9) | (2.7) | (5.4) | (12.1) | (7.4) | (4.5) |

* Degrees Celsius denoted by brackets; standard deviations computed only in °F. For comparison, degree days for this period were fall 15.5, winter 41.8, spring 12.5, computed on 65°F base. NA denotes data not available (less than six weeks data).

** Includes only 1 April - 16 May 1984.

seven houses is small and not necessarily representative of any larger population. We regard this project as a deep and comprehensive study of case histories that will yield hypotheses that will be tested on a larger sample.

## SEASONAL CHANGES IN THERMOSTAT SETTINGS

Table 1 provides an overview of our thermostat data during the past heating season (1 July 1983 through 15 June 1984). The table is divided into fall (1 September through 30 November), winter (1 December through 31 March), and spring (1 April through 15 June). We selected dates for the beginning of the fall season and the end of the spring season so as to include all active use of the thermostat. As the weather grew colder in fall, a consistent pattern of thermostat adjustment developed in most houses. Winter dates were selected to bracket the months when this pattern was most consistent for the greatest number of houses. Thus our seasons are delimited by behavioral rather than meteorological criteria. Mean season figures are computed only for houses with at least six weeks data in that season.[2]

Since the thermostat only activates the furnace when needed, theoretically it should not be necessary to turn it down during warm spells. We found, however, that none of the households took this approach. Table 1 shows that six out of seven of the houses exhibit a higher mean thermostat setting during winter than during fall or spring. The one apparent exception, House 2, many not be an exception since the calculated thermostat setting is based upon only the first six weeks of the spring period, when ambient temperatures are lower and thermostat settings are probably higher than in the last five weeks of this period. The table also illustrates the range of winter means in our sample, from 64°F (18°C) for House 4 to 74°F (23°C) for House 1. The standard deviation of thermostat readings is lowest in the winter for five of the seven houses. In sum, Table 1 shows that in the winter, as compared to fall and spring, most households maintain a higher average setting and adjust their thermostats through a narrower range.

Figure 1 illustrates the differences between the seasons in graphic form. It plots the daily mean thermostat setting at House 1 against the daily mean external temperatures recorded at that house. The contrast between winter month and spring month settings is particularly marked at this house. Lower settings during the second and third week of February correspond to record high external temperatures. On the other hand, the household did not raise its settings in response to record low temperatures in December and January. This house made the fewest winter thermostat adjustments of any in our sample, but all houses show a similar pattern of more modest winter adjustments and more extreme and more irregular changes during the fall and spring. Figure 1 reinforces the conclusion drawn from Table 1—that people frequently adjust the thermostat according to outside conditions rather than letting it operate automatically.

Since most households exhibit clear seasonal differences in thermostat setting patterns, we will investigate the transition between fall and winter in greater detail. We will use House 4, which has a consistent night setback pattern throughout the winter. Figures 2 and 3 show six weeks of data for House 4 between 25 September

---

[2] Winter data include 1 December - 31 March, except as follows: House 3: 1 December - 19 February; House 4: 1-3 December, 5-30 December, 11-17 and 27-31 January, 22 February - 31 March; House 5: 2 December - 31 March; House 6: 1 December - 3 January, 4 February - 31 March; House 7: 1 December - 3 January, 14 - 31 March.

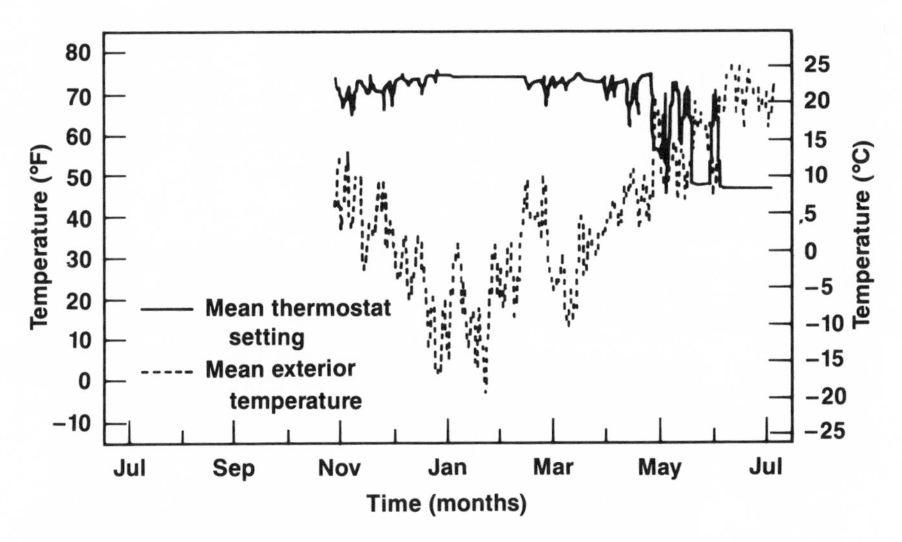


**Fig. 1. Daily mean thermostat settings at House 1, 1983–84.**

and 5 November 1984, including interior temperature, exterior temperature, furnace run-time, and fifteen-minute thermostat settings. During the earlier part of the fall, seen in Figure 2, the thermostat setting is changed only three times. In the interview, this family noted that they tried to delay turning up the thermostat during the fall, bearing with some cold weather if necessary. Figure 2 demonstrates this effect during the period of 25 September through 1 October. The outside temperature never rises above 55°F (13°C), the inside temperature falls as low as 58°F (14°C), yet the household keeps the thermostat at 48°F (8.9°C). It is finally raised to 58°F (14°C) on 2 October, where it is kept until 5 October. In our final interview with this family, we pointed out that this 58°F (14°C) setting was not high enough to cause the furnace to come on. They said they wanted it at this low setting as "insurance," in case one night was especially cold. On the evening of 20 October, the thermostat was briefly (one 15 minute reading) set up to 60°F (16°C), then it was lowered to 55°F (13°C), where it remained until 25 October. We hypothesize that someone decided to make sure that the furnace was operating, which they did by raising the thermostat setting until they could hear the furnace blower (notice the associated blip in furnace run-time).

Figure 3 shows periods when thermostat use is similar to the winter pattern (regular night setback with some daytime setback)

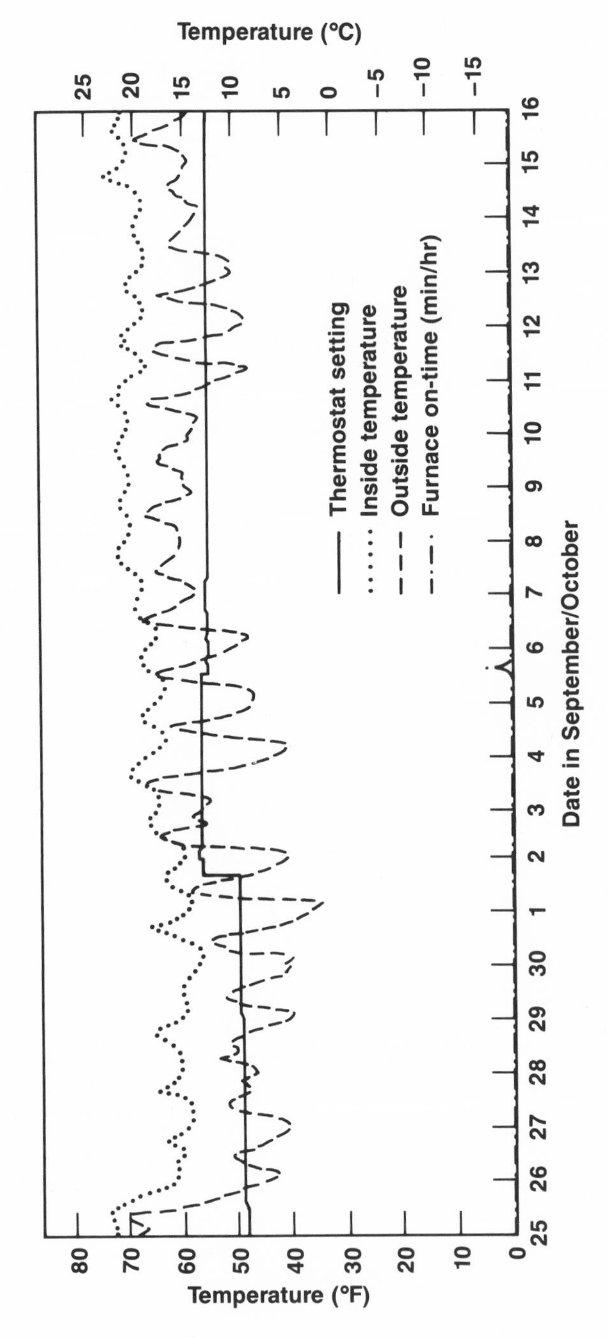


Fig. 2. Thermostat settings at House 4, 1984.

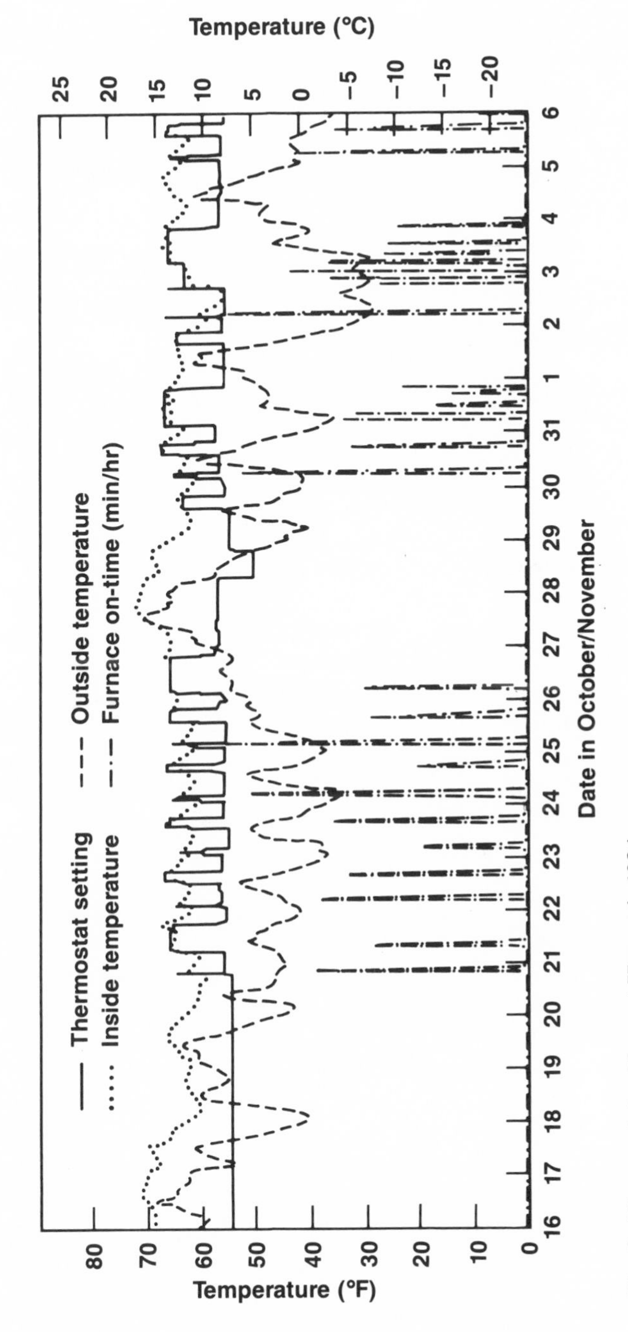


Fig. 3. Thermostat settings at House 4, 1984.

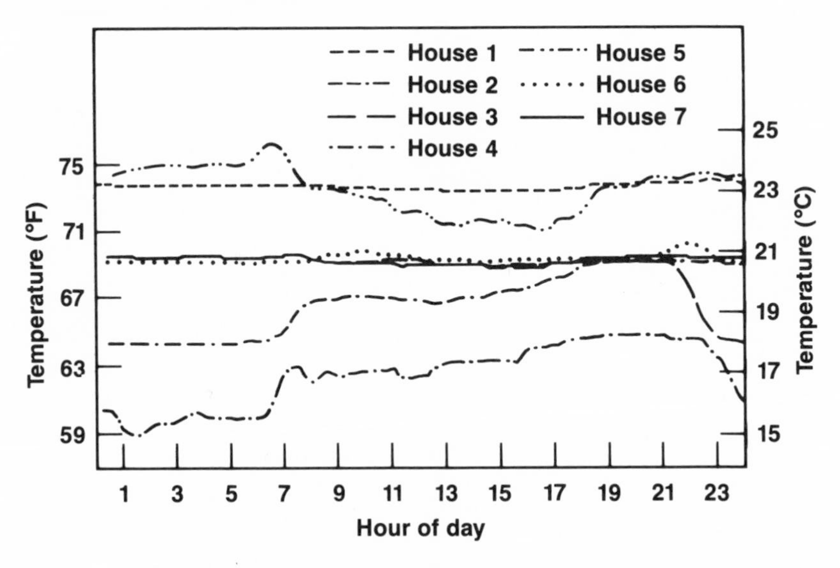


**Fig. 4. Average thermostat settings during winter for all seven houses, 1983–84..**

interspersed with periods when the thermostat is managed quite differently. Out of the 21 days shown there are six days with single nighttime setback and six days exhibiting day and nighttime setback. The thermostat is kept at 55°F for four days leading into the first appearance of the winter pattern. On 28 and 29 October, the exterior temperature was quite warm, causing the household to abandon the winter pattern of household management. On 29 October interior temperature was well above normal, and the thermostat was lowered to 50°F (10°C), well below the normal range. This event might represent the belief that the house would cool down faster if the thermostat setting were lower. After four days of the winter pattern, we see three more days of anomalous behavior (2-4 November). The short spikes that occur in the thermostat settings on the morning and evening of 2 November suggest that someone in this household may have a valve theory of furnace operation. A person with a valve theory (Kempton 1986) believes that the house will warm up faster when the thermostat is set higher. The figure shows that the thermostat was turned up long enough to raise the interior temperature, but was dropped before the interior temperature got even close to the high setting. In the interviews, one of the adults did report a valve theory of thermostat operation. Since the adult with the valve theory was *not* the adult who usually adjusts the thermostat, we see these valve-theory spikes only rarely.

## DAILY THERMOSTAT PATTERNS DURING WINTER

In discussing daily patterns, this chapter will concentrate on data from the winter season, when daily patterns show the most stability. In the seven house sample we found that three or four houses have a cyclic daily pattern. Figure 4 plots average thermostat settings during the first winter of the study, for each 15 minute period of the day. Houses 1, 3, 6, and 7 appear to have stable thermostat settings throughout the day. The lines for Houses 3, 6, and 7 overlap, since they have virtually the same mean setting—just below what one informant called "the big seven-oh" (70°F). House 6 actually has a small night setback, 3°F (1.7°C), which became visible when we replaced the original faulty thermostat sensor. Houses 2 and 4 exhibit night setback, while House 5 shows a daytime setback.

Table 2 compares the daily thermostat patterns of the seven houses during winter. We have developed several numerical measures to capture these daily patterns. We define a "change" as a movement of at least 1°F (0.55°C) between readings. This criterion eliminates spurious changes due to mechanical expansion or vibration as well as instrument drift, which is on the order of ± 0.2 to 0.6°F (0.11 to 0.33°C) over a 24-hour period. Three houses change their thermostats less than once a day; Houses 1 and 3 average only one change every three days. Two houses (2 and 4) change their thermostat about 2.5 times a day. Both of the houses practice consistent night setback and less consistent daytime setbacks.

House 5, which has the highest mean number of changes, practices daytime setback but also changes the thermostat frequently on an eccentric basis. This can be seen in Figure 5, which plots thermostat settings over a two week period in December. Settings for this period range from 50° to 85°F (10° to 29°C), sometimes changing by 25°F (14°C) within 15 minutes. The head of the household reported that the thermostat was malfunctioning, but we could find no evidence for it when we inspected both the system and our instrument data, nor could any problem be found by an independent heating service company.

The daily average size of change is computed for each day in which one or more changes occurred. Table 2 shows that the smallest changes occur in households that least frequently make changes: Houses 1, 3, and 6. These houses rarely change their thermostat and make only slight adjustments when they do change it.

**Table 2. Daily Patterns of Thermostat Settings, Winter 1983-84 and Spring 1984.**

| | House | | | | | | |
|---|---|---|---|---|---|---|---|
| | 1 | 2 | 3 | 4 | 5 | 6 | 7 |
| *Number of Changes per day* | | | | | | | |
| Winter | 0.4 | 2.4 | 0.3 | 2.6 | 4.5 | 1.7 | 0.7 |
| Spring | 0.8 | 2.4 | 0.3 | 1.1 | 2.7 | 2.2 | 0.2 |
| *Mean Setting °F [°C]* | | | | | | | |
| Winter | 73.6 [23.1] | 66.5 [19.2] | 69.2* [20.7] | 63.6* [17.6] | 73.5 [23.1] | 69.5* [20.8] | 69.2* [20.7] |
| Spring | 60.2 [15.7] | 66.9** [19.4] | 64.2 [17.9] | 59.2 [15.1] | 60.4 [15.8] | 65.0 [18.3] | 68.4 [20.2] |
| *Mean change °F [°C]* | | | | | | | |
| Winter | 4.3 [2.4] | 5.6 [3.1] | 2.4 [1.3] | 7.2 [4.0] | 8.6 [4.8] | 2.6 [1.4] | 5.6 [3.1] |
| Spring | 11.1 [6.2] | 5.3 [2.9] | 3.1 [1.7] | 8.2 [4.6] | 18.0 [10.0] | 6.2 [3.4] | 5.0 [2.8] |
| *Daily high °F [°C]* | | | | | | | |
| Winter | 74.2 [23.4] | 70.4 [21.3] | 69.5 [20.8] | 67.0 [19.4] | 80.1 [26.7] | 71.5 [21.9] | 70.7 [21.5] |
| Spring | 63.2 [17.3] | 70.3 [21.3] | 64.7 [18.2] | 62.0 [16.7] | 72.4 [22.4] | 68.9 [20.5] | 68.9 [20.5] |
| *Daily low °F [°C]* | | | | | | | |
| Winter | 72.9 [32.7] | 63.4 [17.4] | 68.7 [20.4] | 58.6 [14.8] | 65.6 [18.7] | 68.0 [20.0] | 67.7 [19.8] |
| Spring | 57.4 [14.1] | 63.4 [17.4] | 63.7 [17.6] | 56.8 [13.8] | 53.1 [11.7] | 61.4 [16.3] | 67.9 [19.9] |

* Data is incomplete. See footnote 2 for dates included.

** Includes only 1 April - 16 May 1984.

House 7 adjusts their thermostat infrequently but by relatively large amounts. (They reported frequent weekend trips.) The night-time setback houses, 2 and 4, also exhibit large changes. House 5, as seen in the sample weeks of Figure 5, changes the thermostat most frequently and by the largest amounts. Predictably, the houses with the lowest average number of changes and size of change also have the least difference between mean daily lows and mean daily highs.

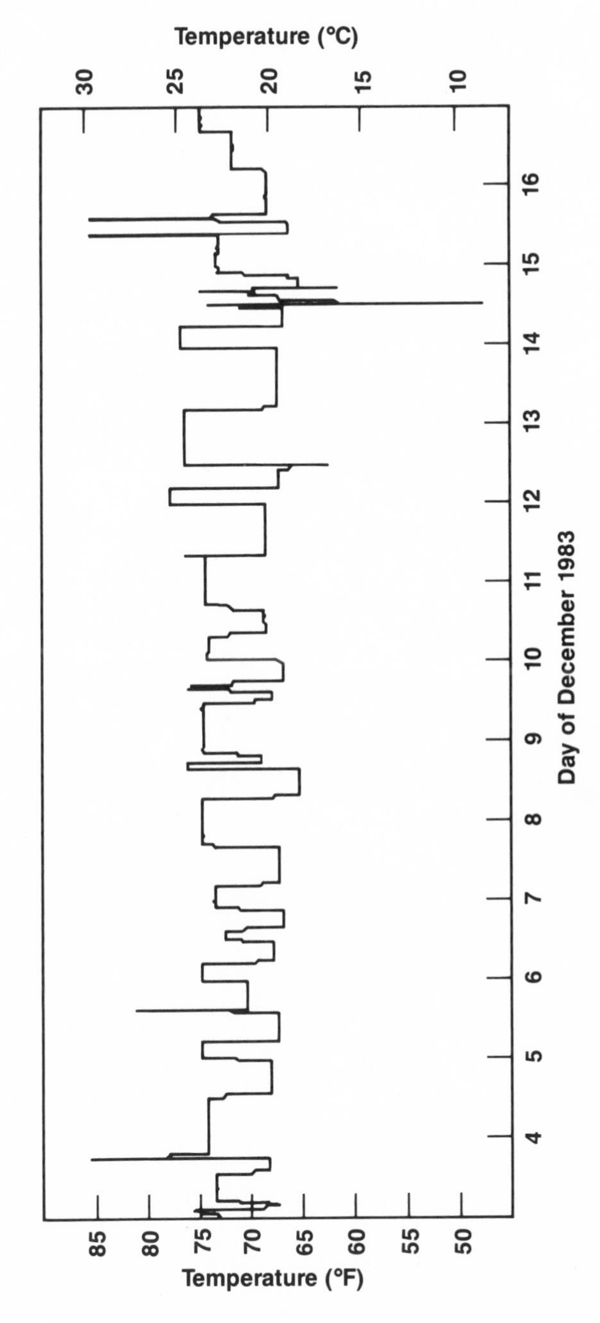


**Fig. 5. Thermostat settings at House 5, 3–17 December 1983.**

Overall, we find three types of daily winter patterns. Three households have regular night setbacks, and less regular daytime setbacks. One of these three makes very small setbacks. Two houses make changes rarely, and only adjust a few degrees when they do change. One house has a highly erratic pattern with frequent and large changes. For discussion of the relation of thermostat settings and daily household schedule, see Weihl (this volume).

## HOUSEHOLD MANAGEMENT

To help interpret our recorded instrument data, we asked participating families about thermostat settings in an interview. This information was solicited in two forms. A written questionnaire was given to each household member at the beginning of the interview. Members were discouraged from consulting each other as they completed the questionnaire. After it was completed, an open-ended ethnographic interview was conducted with adult family members, and any others who reported setting the thermostat.

Households were able to give a clearer account of their thermostat management strategies in the interviews than on the questionnaire. With the exception of a few specific questions, discussed below, most households were able to accurately describe their thermostat management strategies. One example of reported management strategy follows. House 1 said that they did not use their thermostat during summer. During the fall and spring they move it up and down some to maintain a comfortable temperature, but they note that on sunny days their house is warm enough and they only need heat for a few hours in the evening. During the winter they report that they only change the setting occasionally. The pattern (except for summer behavior) is confirmed by Figure 1 and our other records.

Ethnographic interviews provide an opportunity for informants to discuss their energy management schemes at length and in as much detail as seems relevant. Many informants first state a policy that represents the ideal, then go on to relate the various circumstances that modify the ideal. Sometimes this revision process ends up contradicting the initial statement, as is shown by this excerpt from an interview with Melanie and Greg in House 3. On the questionnaire Greg wrote the following thermostat schedule: 66°F (19°C) when at work, 68°-70°F (20°-21°C) when at home in the evening, and 68°F (20°C) at night. He explained this later in the interview:

*Greg:* Well you see a lot of time when I go to bed I will turn it down just a tiny bit and because I don't mind it being cool when I get up.

*Interviewer:* Okay that will be just before getting into the bed.

*Greg:* (agrees)

*Interviewer:* Now I am trying to get a rough percentage estimate of how often you'd say you turn it down when you go to bed versus leaving it there.

*Greg:* Gee I don't know.

*Interviewer:* Just real rough like 90% or half the time or occasionally.

*Melanie:* I think about it once a year that's about it when I go to bed. I think about it once a year.

*Greg:* Well probably wouldn't be—just occasionally because I usually don't think about it too much, you know. If it feels a little warm I'd get up and turn it down a little bit.

*Interviewer:* Like after you go to bed it might feel a—too warm.

*Greg:* Yeah, I've gotten up through the night and turned it down, yeah—very seldom. Very seldom.

As seen in Figure 4 and Table 2, Greg's "very seldom" answer is much more accurate than his original claim.

Respondents more typically amend statements of ideal policy in less drastic ways. A variety of circumstances are mentioned as occasions for deviation from normal behavior. All of the houses except one indicated that they turn down the thermostat when they are away from home for extended periods of time, and our instrument data verify this practice. The household that was the exception reported that someone was always at home. Some residents turn the thermostat down when they will be out of the house for a few hours, but they apparently try to judge whether they will be gone long enough to make lowering the thermostat worthwhile. For example, Renee from House 2 says:

*Renee:* But I don't, if I am going to be gone an hour or two, I usually leave it up, I don't turn it down. If I am going to be gone for the afternoon or for more than two hours then I turn it down, but I figure an hour or two hours won't make that much difference.

Informants also report that they turn the thermostat down or fail to turn it up to its normal daytime setting during warm, sunny days. This effect can be observed on Figure 1, in which warm, sunny weather during the second and third week of February correspond to lower thermostat settings.

Some circumstances are believed to require higher than normal thermostat settings. Two houses mentioned that they set the thermostat higher during illnesses. House guests also alter normal behavior patterns. Rooms that are normally closed off are opened up. The thermostat setting may be higher or lower in accordance with the guest's comfort. In some cases this is the subject of circumspect negotiation, as revealed by the following reference to blankets in the living room at House 2:

*Renee:* On the other hand, when we have guests, I always point out to them that (pause) on every CHAIR, there's something to cover up with (laughs).

*Bill:* Yeah, but it does (pause) uh, make you feel a little chintzy you know, and so . . .

*Renee:* Well I don't do that with everybody, but our good friends, I do, and they say "Oh, well we do the same thing."

Households also report that they vary their thermostat settings according to activities. Sedentary activities, like reading, require higher temperatures. Physical exercise or heat generating activities, like cooking, may require lower temperatures.

Informants noted a variety of alternatives to turning up the thermostat when they felt cold. All of the families mentioned that they put on more or warmer clothing when they felt cold. Many houses keep blankets and other coverings accessible to provide warmth during inactive periods. Three houses have some rooms that are noticeably warmer than others, and household members sit in those rooms when they want to be a little warmer. Other strategies included doing some cooking to warm up the kitchen, taking a warm bath, or using auxiliary heating. Many respondents mention electric blankets or flannel sheets on their bed, or the use of a heated water bed to enable them to turn the thermostat down at night. This information aids in interpreting the instrument data, for it explains how families can tolerate colder temperatures without raising the thermostat.

Five of the seven houses reported household discussions of energy management and sometimes found it the subject of debate. These households all had clear ideas as to who should set the thermostat and the circumstances under which changing the setting was appropriate. In three of the households one family member seemed to set policy, subject to the agreement of other family members. For example, in House 4, Nick was a major force in attempting to maintain 67°-68°F (19.4°-20°C) as the evening setting and 55°-60°F (13°-15.5°C) as the night setting. This is seen in the following discussion with his wife:

*Interviewer:* But do you, do you argue about it? I don't . . .

*Ellen:* Sometimes, sure.

*Interviewer:* You do argue about it?

*Ellen:* Oh, sure.

*Interviewer:* What, in what ways?

*Ellen:* I WANNA TURN IT UP NICK! (laugh) No, I don't know. It's not that bad. Sometimes. We talk about it.

*Interviewer:* Like "turn it up it's too cold in here?"

*Nick:* And I usually wimp out.

*Ellen:* Yeah, he does.

*Interviewer:* He figures it's not worth the fight.

*Nick:* It's not worth sleeping alone. (everyone laughs) It's real *cold* in this house—you lose two ways.

Some of the households discussed how these policies were enforced, but only one felt that enforcement was worth major discord.

## REPORTED VS. OBSERVED THERMOSTAT SETTINGS

Because we collected data from a variety of sources, we were able to compare the answers that might be obtained on a typical survey to answers obtained during intensive interviewing and to our instrument records. Our comparisons must be interpreted in the context of the sample's self-awareness: Since the households were necessarily aware that their thermostat settings were being monitored, one might expect them to be less likely to bend the truth in

**Table 3. Reported and Observed Thermostat Settings, Winter 1983-84**

| | House | | | | | | | |
|---|---|---|---|---|---|---|---|---|
| | 1 | 2 | 3 | 4 | 5 | 6 | 7 | all |
| Written on Questionnaire | NA* | 65.1 | 67.2 | 60.7 | NA* | 66.4 | 70.0 | 65.9 |
| Reported in Interview | 71.0 | 66.4 | 68.4 | 60.5 | 70.9** | 66.6 | 70.0 | 67.7 |
| Observed | 73.6 | 66.5 | 69.2 | 63.6 | 73.5 | 69.5 | 71.6† | 69.6 |
| **Difference** | | | | | | | | |
| Interviews minus Observations | -2.6 | -0.1 | -0.8 | -3.1 | -2.4 | -2.9 | -1.6 | -1.9 |

* Not available (NA) indicates that no thermostat setting was filled in on questionnaire.

** Reported setting for House 5 is approximate, based on unclear statements during interview.

† Two periods when thermostat was kept below 55°F for longer than 24 hours were deleted from calculations.

order to impress us with their energy-efficient behavior. Table 3 compares reported and actual thermostat settings during the winter of 1983-84. The table condenses a great deal of information; a more complete discussion of reported and actual thermostat settings will appear in a subsequent paper by Krabacher and Kempton.

The reported settings for both interviews are computed as an average weighted by the duration of the setting.[3] Therefore, inaccuracies in the reported average setting may be due to inaccuracies in the reported setpoint, in the reported duration of the setpoint, or in the regularity with which the schedule is followed. The mean setting reported in the questionnaire was 2.2°F (1.2°C) below the observed thermostat setting. The mean setting reported in the interview was 1.9°F (1.1°C) below the observed setting.[4]

In sum, even though the informants knew their thermostat set-

[3] For example, if the informant said that the thermostat was set at 70° from 8 am to midnight and 65° from midnight to 8 am, we would compute the weighted mean as: (16 hours x 70° + 8 hours x 65°) / 24 hours = 68.3°.

[4] These figures are slightly lower than those in the conference draft of this chapter due to our using more conservative analysis that minimized differences between reported and observed means.

tings were being recorded, we see a consistent under-reporting of thermostat setting. Although the sample is tiny and the data are imperfect, we tentatively conclude that reported thermostat settings derived from surveys should be adjusted upward by at least 2°F (1°C) to estimate actual mean thermostat settings for the sample. We say "at least" because the discrepancy could be larger if participants did not know that their answers were verifiable. Unfortunately, we do not yet have sufficient data to estimate which families are likely to under-report by larger amounts; in our data, underreporting ranged from 0.1°F-3.1°F. Thus, while we provide a minimum figure for adjustment of sample means, we cannot yet provide guidance in reducing the variance, which would improve correlation studies.

## SUMMARY

In this small sample of seven houses, we find great diversity in both average winter thermostat settings (from 64° to 74°F, or 18° to 23°C) and in patterns of daily thermostat changes. Three of the study houses followed a consistent nighttime setback. Two maintained a steady setting, with changes made infrequently and only by small amounts when they did occur. One house made frequent, large, and unpredictable changes throughout the day. For six houses, winter settings are predictable, while the spring and fall thermostat setting is more erratic and more determined by outside temperature. None of the households let the thermostat operate automatically during the fall and spring.

Most families reported that one member was primarily responsible for managing the thermostat. They report alternative strategies for keeping warm, such as wearing more clothing, covering oneself with blankets, or moving to a warmer room in the house, all of which permit lower thermostat settings. Despite the fact that these households knew their thermostat settings were being recorded, the thermostat settings they reported in interviews were consistently below the observed settings.

### Acknowledgements

*This work is supported by grants from the National Science Foundation (BNS 82-10088), the MSU Agricultural Experiment Station (Project 3152), and the Kellogg Biological Station Small Farms Project. It is part of the Family Energy Project, which is housed in the Institute for Family and Child Study, College of Human Ecology, MSU.*

**References**

Diamond, R.C. "Energy and Housing for the Elderly: A Closer Look." This volume.

Erickson, R.J. "Household Energy Use in Sweden and Minnesota: Image and Reality." In *Families and Energy*, ed. B. Morrison and W. Kempton. East Lansing, MI: Institute for Family and Child Study, College of Human Ecology, Michigan State University, 1984.

Fels, M.F., and M.L. Goldberg. "With Just Billing and Weather Data, Can One Separate Lower Thermostat Settings from Extra Insulation?" In *Families and Energy*, 1984, (see above).

Kempton, W. "Two Theories of Home Heat Control." *Cognitive Science* 10:75-90, 1986.

Kempton, W., C.K. Harris, J.G. Keith, and J.S. Weihl, "Do Consumers Know 'What Works' in Energy Conservation?" *Marriage and Family Review* 9:115-134, 1985.

Luyben, P.D. "Prompting Thermostat Setting Behavior: Public Response to a Presidential Appeal for Conservation." *Environment and Behavior*, 14:113-128, 1982.

Vine, E. "Saving Energy the Easy Way: An Analysis of Thermostat Management." Draft report available from author. Lawrence Berkeley Laboratory, Berkeley, CA, July 1984.

Weihl, J.S. "Family Schedules and Energy Consumption Behaviors." This volume.

Wilk, R.R. and Wilhite, H.L. "Why Don't People Weatherize Their Homes?: An Ethnographic Solution." This volume.

# Family Schedules and Energy Consumption Behavior

*Jeffrey S. Weihl*
*Michigan State University*

## INTRODUCTION

The factors influencing residential energy consumption can be divided into two domains, structural characteristics of residences and occupant behavior. It has been hypothesized that occupant behavior accounts for a large proportion of energy-use variation. One well-known study, conducted by Princeton at Twin Rivers, estimated that 38% of the variation of heating energy consumption among its sample of identical townhouses was due to occupant consumption patterns (Socolow 1978). Another study found a greater that twenty-to-one variation in consumption among identical units in an apartment building (Lipschutz et al. 1984).

Despite their importance, behavioral variables have received considerably less attention and are less well understood than structural variables. One reason for this lack of research is the complexity of human behavior and the difficulty in measuring it. While behavior is complex, it does follow patterns. This chapter focuses on these patterns and investigates the hypothesis that many energy-related behaviors in residences are strongly influenced by the daily schedules of the occupants.

This chapter utilizes data gathered from four occupied residences. It focuses on the occurrence and patterning of measur-

*Energy Efficiency: Perspectives on Individual Behavior*

able behavior rather than on the effects of this behavior on energy units consumed. For that reason the data presented in this chapter deal with patterns of thermostat *adjustments* and hot water use *events* rather than with the exact values of thermostat settings and hot water consumption by volume. (For the latter type of analysis see Kempton, this volume, and Kempton and Krabacher, this volume.)

## DISCUSSION

Energy behaviors may be the result of conscious management or unconscious habit. Behaviors may also be infrequent and nonsystematic or frequent and regular. Social scientists studying energy behavior are interested in all these types of behavior, yet recognize that the behaviors that are most easily studied and that probably have the greatest impact on energy consumption are frequent and show regularity. The identification of patterns (regularity over time) and the reasons for them is an important task in the behavioral study of energy consumption.

Energy behaviors may be patterned seasonally, monthly, annually, or daily. This chapter focuses on daily patterns of energy behavior to provide some preliminary estimates of how consistent energy behaviors are. While studying all types of patterns is important, daily behaviors take place most frequently. Thus examining the effects of daily family schedules on energy behavior is a logical beginning.

The behaviors affecting household energy consumption are numerous and have effects of different magnitudes. It would be impossible to study the patterns of all behaviors that influence energy consumption. Therefore, this chapter focuses on behaviors that most directly affect the two major energy-consuming systems within most residences, the space-heating and water-heating systems. Of the many behaviors affecting the performance of these systems, this chapter looks only at thermostat setting and use of hot water.

We have two goals in identifying patterns of energy-related behaviors in these case studies. First, we hope to identify variables that might affect patterning of energy behaviors for further study. Second, we believe that identifying behaviors significantly affected by daily patterns could help policymakers to target those uses of energy that could be better managed by utilizing these patterns. For instance, if some water uses are greatly influenced by the daily routine of a household, point-of-use hot water heaters at those taps might conserve significant amounts of energy. Similarly, the use of clock thermostats (perhaps for water heaters in addition to those for furnaces) might significantly reduce energy consumption in households with very regular thermostat setting patterns.

## DATA SOURCES

The data presented in this chapter were gathered in the winter of 1983-84 from households in the Lansing, Michigan area. The sample times for water use and thermostat settings vary due to mechanical constraints; they average 43 days in length. Hourly thermostat settings were collected by instruments and are used to determine patterns of thermostat adjustment. Instruments also report the time of day and point of use of each "hot water use event."[1]

## THE FAMILIES AND THEIR SCHEDULES: INTERVIEW AND INSTRUMENT DATA COMPARED

This section compares the reported family daily schedules with parallel instrument data to examine daily behavior patterns. When behaviors are described as not being patterned we mean that no *daily* pattern was apparent, not that such behaviors are necessarily random or haphazard. Median hourly thermostat settings are utilized to construct "typical" daily thermostat setting patterns.[2] Hot water use events at each tap were grouped hourly using a 24 bin histogram to determine the relative frequency of events occurring at different hours of the day. We used data for each house to compare the daily schedule with the measured use of the thermostat, and with the amount of hot water used in kitchen, bathroom sinks,

---

[1] A hot water use event is defined as any consumption of hot water over 0.1 liter/min., for which the point of use is identified. (The monitoring system is discussed in detail in Weihl et al. 1984; Weihl and Kempton 1985). Family schedules were elicited from each household as part of broader ethnographic interviews. These interviews asked all family members present to report their regularly scheduled weekday and weekend times for various energy behaviors.

[2] Median values are used rather than means, since infrequent extreme settings and other factors tend to obscure the pattern of a "typical" day. Since this analysis is more concerned with patterning in habitual thermostat raising or setback than with the actual value of thermostat settings, the use of medians is justified. For a more detailed examination of actual thermostat setting values over time see Kempton and Krabacher, this volume.

bathing and laundry.

We compared schedules and instrument data for five families. In the interest of clarity, however only four families are presented in detail in this chapter. (Family 3 did not exhibit sufficiently different behavior patterns to warrant a detailed presentation here.)

**Family 1**

Family 1 is composed of a retired industrial worker, Eduardo (62), his wife, Flora (58), and their children living at home, Willy (21), Robert (20), and Anna (19). Robert and Anna work during the day, Willy works at various times, often at night. Two of their granddaughters stay at the house after school. Eduardo and Flora do not often leave the house. Flora spends much of the day in the kitchen cooking, and many family members arrive during the day to visit and eat lunch. Consequently the kitchen and dining room see almost constant activity throughout the day.

| Monday-Saturday | | | |
|---|---|---|---|
| 200-300 | Eduardo to bed | 1200 | Willy to work |
| 600 | Flora gets up, uses sink, showers | 1200-pm | Flora uses kitchen sink |
| 715 | Robert gets up, showers | 1200-1230 | Lunch time |
| 800-830 | Anna gets up, showers | 1800 | Eduardo showers (2X week) |
| 900-1000 | Eduardo gets up uses sink | | |
| 1000 | Willy gets up | 2300-2330 | Flora to bed |

On Sundays, Eduardo and Flora go to church early and get home in the late morning. Flora cooks a large meal, which the family (including those who live elsewhere in town) gather to eat around noon.

**Thermostat Use.** Family 1 reports that they rarely change their thermostat settings. Since Eduardo and Flora are always at home and the house is the setting for a great deal of activity during the day, a daytime thermostat setback pattern would not be practical. They report no nighttime thermostat setback pattern. The data show that the median thermostat setting was a constant 74°F (23.3°C). Thus, in Family 1, thermostat setting behavior is not related to the occupants' daily schedule.

**Kitchen Sink Use.** Since Flora cooks and works much of the day, we expect that the kitchen sink would be used all day and especially between 1200 and 1900. We would not expect this pattern to vary much during the weekend. As expected, kitchen sink use does start between 600 and 700 each day and continues to

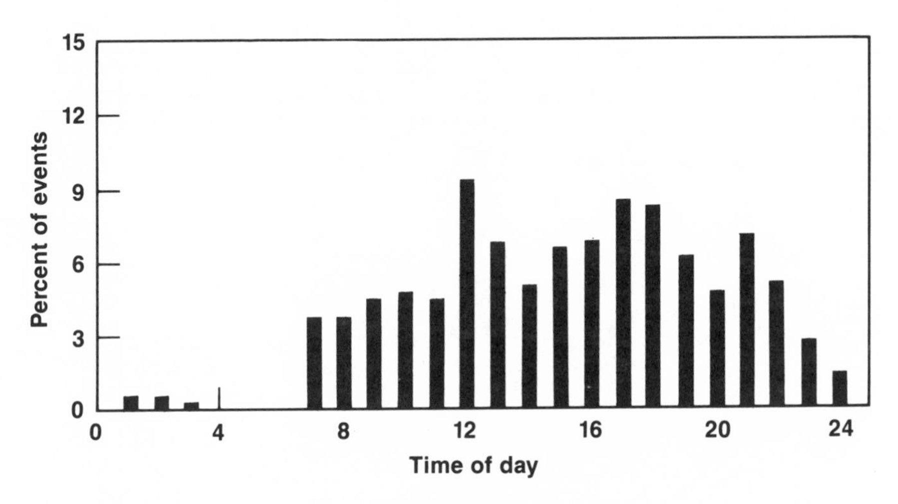


**Fig. 1. Kitchen sink use events, Family 1 (Monday through Friday).**

become more frequent until noon, with another peak around dinner, and remains frequent until 2200 (Figure 1). Weekend kitchen tap use is similar but with a greater proportion of events taking place between 1000 and 1100, possibly due to preparation for the Sunday family meal (Figure 1 is a good illustration of a behavior that is not consistently scheduled at a few specific times of day.)

**Bathroom Sink Use.** We would expect to find most bathroom sink events occurring in the periods 600-700 and 900-1000. In fact, bathroom sink use occurs frequently throughout the day and is most frequent from 900-1000. Other than this peak, and another around dinner time (1600-1700), bathroom sink use is quite scattered and does not well reflect the family's daily schedule.

**Shower/Tub Use.** Most of these events would be expected at 600, 715-830 and 1800. Since 15% of events at the upstairs shower and 31% of events at the downstairs shower occur between 1800 and 1900, a pattern of shower use is indicated, but this pattern conflicts with the one reported by the family.

**Laundry Use.** No laundry pattern was reported by the family or shown in the data. Thus laundry behavior is affected little by daily schedule.

**Family 2**

Family 2 is a two-person household composed of Bill (62) and his wife Renee (58). Bill works Monday through Friday, Renee does volunteer work on a flexible schedule Monday through Friday. They report that their schedule is regular, especially with regards to thermostat settings and bathing behavior.

| Monday-Friday | | | |
|---|---|---|---|
| 620 | Bill gets up | 1300 | Renee leaves (2X week) |
| 630 | Bill uses sink, showers | 1600-1700 | Renee home, turns up thermostat |
| 700 | Renee gets up and showers | | |
| 700-715 | Breakfast dishes rinsed | evening | Renee uses kitchen sink |
| 710 | Bill leaves for work | 2200 | Both go to bed |
| 800 | Breakfast dishes are washed | | |
| 900 | Renee leaves (3X week), turns down thermostat | | |

On weekends, there is a variable schedule, especially on Saturday. The occupants report that they often leave the thermostat up (70-72°F) all day on Saturdays and Sundays.

**Thermostat Use.** This family reports both nighttime thermostat setback and weekday daytime thermostat setback. We would expect from their reported schedule to find upward thermostat adjustments at 700 when they get up, at 1600-1700 when they come home, and perhaps at 2100-2200 when they report they often "nudge it up a bit, when they are less active." Downward thermostat adjustment would be expected at 900 and 2200. Instrument data confirms most of this report (Figure 2). The thermostat is normally set up (to 66°F) between 700 and 800, set up (to 67-69°F) between 1600 and 1800, and set back (to 65°F) between 2200 and 2300 as reported. However, the reported daytime setback pattern between 900 and 1600 is not evident. This daytime setback is probably less frequent than the informants suppose, due to Renee's varied daytime schedule of being home some days and not others. A scatter plot of weekday hourly thermostat settings (Figure 3) clearly shows that thermostat settings closely follow their daily schedule and also shows evidence of some daytime setback. We should expect a less rigid thermostat schedule during weekends. Any apparent pattern should show varied settings during the daytime hours with consistent nighttime setback. The data tend to support this prediction; daytime setting is quite variable, yet nighttime setback is clear (Figure 4).

**Kitchen Sink Use.** We would expect kitchen sink uses to cluster around 700-830, 1200, and in the evening (perhaps 1900). The

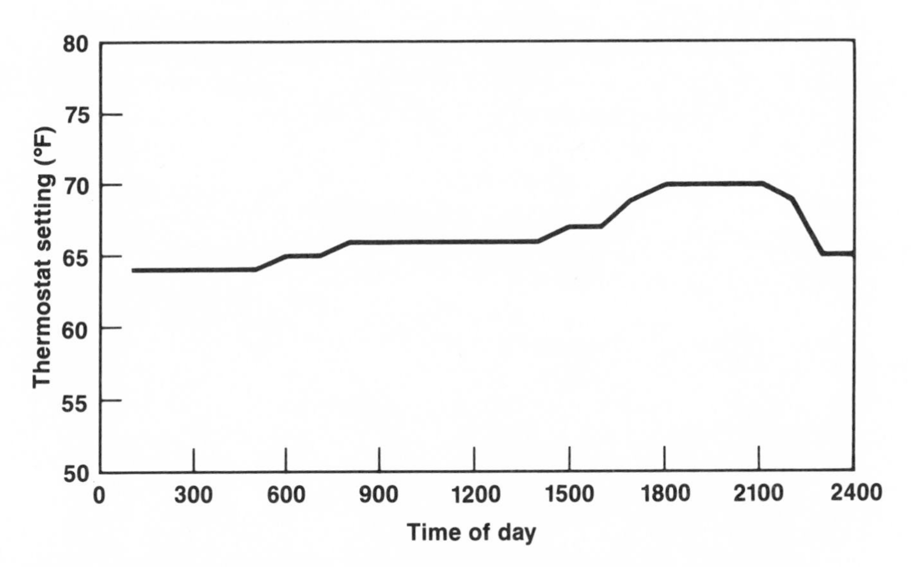


**Fig. 2. Median hourly thermostat settings, Family 2 (Monday through Friday).**

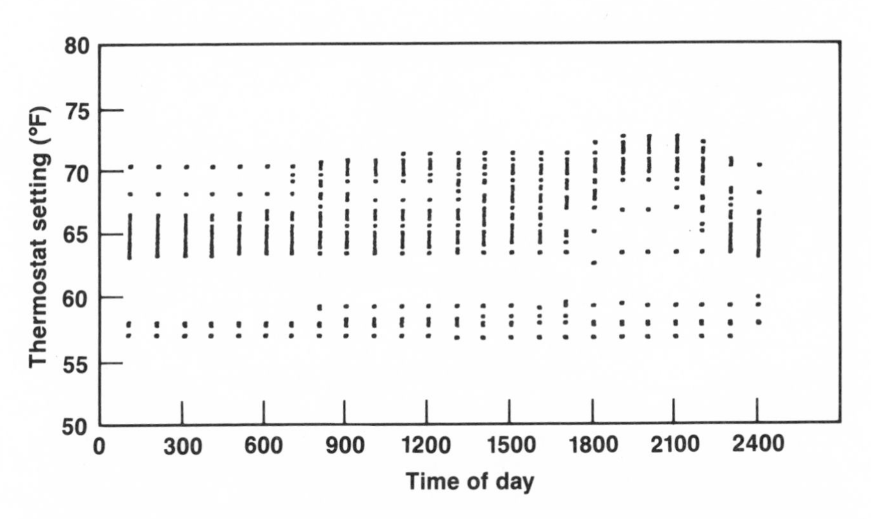


**Fig. 3. Thermostat setting scatterplot, Family 2 (Monday through Friday, hourly readings).**

data indicate that on weekdays, kitchen sink events are highly patterned: 34% of these events take place during 630-800 and 24% during 1800-1900. On weekends another pattern appears: morning kitchen sink use takes place later 900-1100 (11%), and there is more afternoon kitchen sink use, 1200-1400 (33%). The normal evening use, 1700-1900, accounts for 24% of weekend kitchen uses. Thus for Family 2, kitchen sink use is highly patterned and corresponds well with the occupants' daily schedule.

**Bathroom Sink Use.** Bathroom sink use would be expected to occur mainly between 600 and 700 on weekdays. The data support this conclusion, as 38% of all bathroom sink uses occur at this time. The use of this sink also appears to be clustered around dinner time and bedtime. On weekends the pattern is both different and less distinct. Weekend usage is clustered around 800-1000 (25%), and takes place regularly throughout the afternoon and evening (1300-1900, 35%) and at bedtime (2200-2300, 8.8%). Bathroom sink use appears to be well-patterned by daily schedule of Family 2.

**Shower Use.** Showering events would be expected to cluster at 630-700 on weekdays and around 800-900 on weekends. Shower events occur regularly on weekdays, between 600-700 (40%) and between 700-1100 (42%). (Figure 5 is a good example of how a consistently scheduled water use appears graphically.) On weekends, showers typically take place between 800-1000 (48%) with a tendency for more showers at late morning and afternoon times. Showering events reflect family schedules in this case extremely well.

**Laundry Use.** The family reported that they do most laundry during the morning. Laundry use exhibits a pattern but use clusters at afternoon times. On weekdays, 53% of laundry events occurred at 1400-1600, and on weekends 72% of uses occurred at 1300-1500. Laundry is more likely to be done on weekdays than on the weekend. Laundry use for Family 2 is also well patterned, however the pattern conflicts with the reported schedule.

**Family 4**

Family 4 is a four-person household composed of Nick and Ellen, both in their 40s, and their children Lisa (11) and Eric (18). Nick and Ellen are both educators, Eric works part-time and attends high school, and Lisa attends school. They report a fairly regular schedule with the house being empty during the daytime. All leave the house at a specified time, except Nick, who sometimes works at

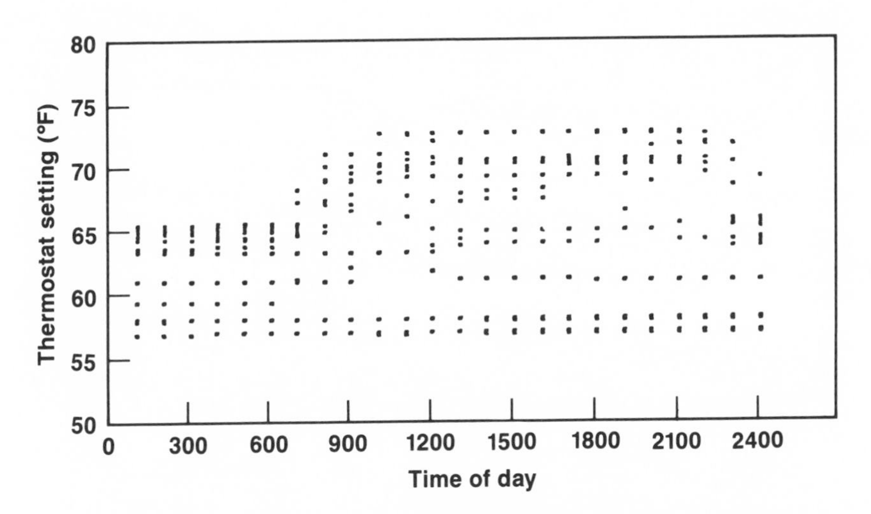


**Fig. 4. Thermostat setting scatterplot, Family 2 (Weekend, hourly readings).**

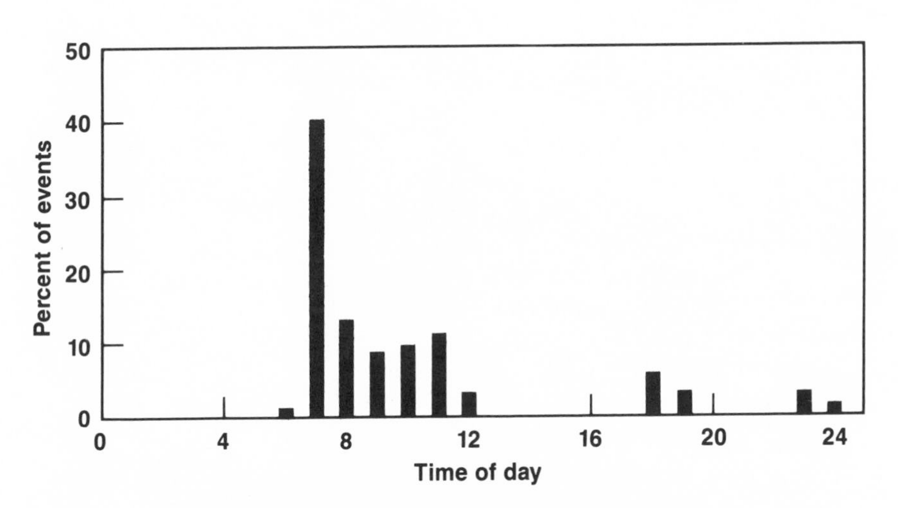


**Fig. 5. Shower use events, Family 2 (Monday through Friday).**

home in the mornings. Weekends are flexible, except for Lisa's fairly regular sleeping schedule.

| Monday-Friday | | | |
|---|---|---|---|
| 600-630 | Ellen gets up, turns up thermostat | 1700 | Nick comes home |
| 630 | Lisa gets up, Nick gets up | 1730-1800 | Eric comes home |
| 630 | Ellen and Lisa shower, use sinks | 1700-1730 | Dinner preparation |
| 700 | Nick uses bathroom sink, showers | 1800-1900 | Dinner |
| 700-715 | Eric gets up and showers | 1900-2000 | Dinner dishes washed |
| 715-745 | All leave | 2030-2100 | Lisa goes to bed |
| 1530-1700 | Ellen comes home | 2200-2300 | Ellen and Nick go to bed |
| 1630 | Lisa comes home | 2300-2400 | Eric goes to bed |

**Thermostat Use.** The residents of this household report that they try to actively manage their thermostat setting to reduce energy consumption. The thermostat is reportedly set up to the mid-to-high 60s (18.3-20.5°C) between 600 and 630, then down to 55°F (12.8°C) when people leave (730-800), and back up to mid-to-high 60s when people come home at 1530-1600 until bedtime (1100) when it is set back to 55°F (12.8°C). Presumably this schedule would be modified on weekends by keeping the thermostat in the 60s during the daytime, since people are home.

Median hourly thermostat settings (Figure 6) show that the thermostat is indeed regularly raised several degrees between 500-700. Daytime setting is a constant 66°F (18.9°C), which does not indicate a daytime thermostat setback pattern. The setting is raised during the evening hours (1800-1900, and set down at bedtime 2200-2400. Weekend thermostat setting (not shown) is much as predicted, showing a constant day and evening temperature (67°F, 19.4°C) and a nighttime setback to 62°F, (16.7°C), with morning set-up occurring later (700-900).

Thermostat setting is visibly patterned in this household, with weekdays and weekends showing patterns that reflect differences in family schedule. The pattern exhibited, however, contradicts the weekday daytime setback schedule reported by the family. A scatter plot of hourly weekday settings (Figure 7) shows that lower daytime settings occur, but their invisibility on the median graph indicates that the daytime setback pattern is less frequent than the family believes.

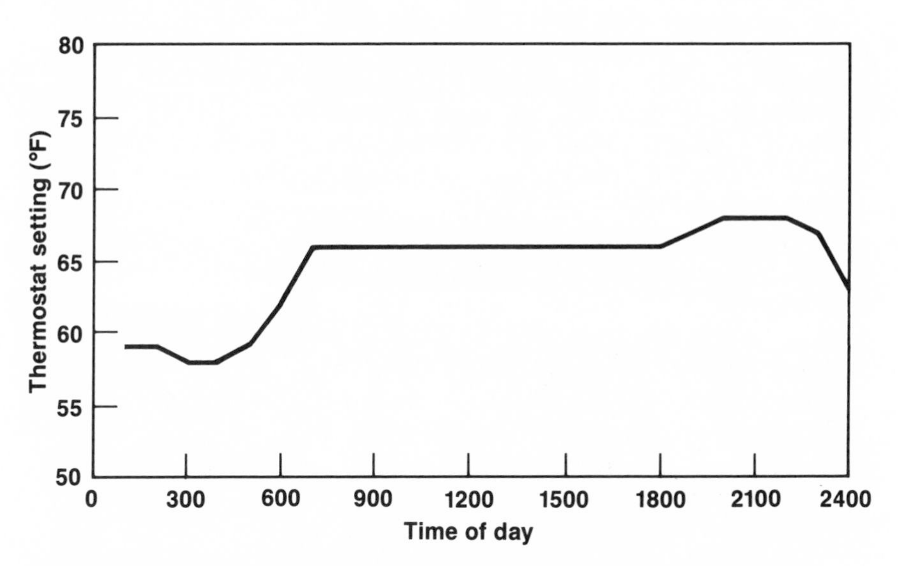


**Fig. 6. Median hourly thermostat settings, Family 4 (Monday through Friday).**

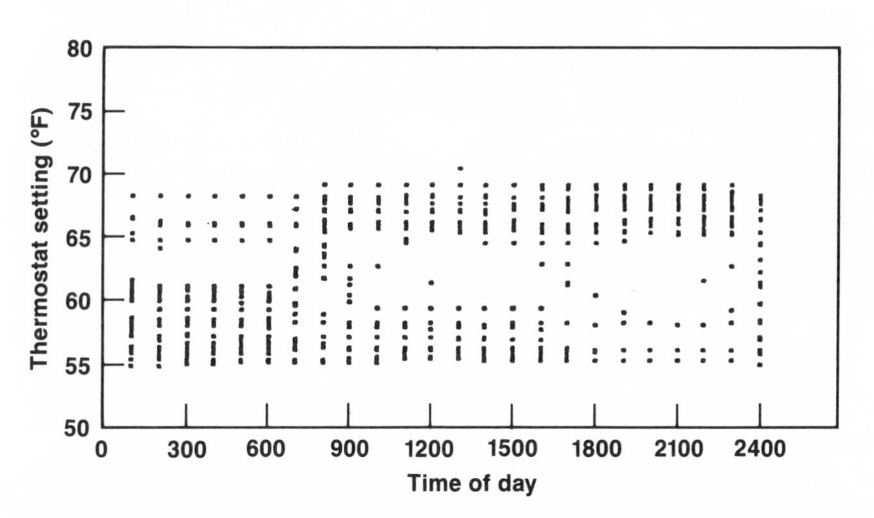


**Fig. 7. Thermostat setting scatterplot, Family 4 (Monday through Friday, hourly readings).**

**Kitchen Sink Use.** Based on information from the interview, we would expect kitchen sink use on weekdays during dinner preparation (1700-1730) and after dinner (1900-2000). The data show that weekday use clusters at 600-900 (16%) and 1700-2100 (49%). Weekday kitchen sink use is highly patterned by the family schedule. Weekend use occurs mainly after 1000, is most intense from 1200-1500 (38%), and continues frequently until 2100. Although weekend usage is less distinctly patterned, it shows that a different family schedule is in effect on Saturdays and Sundays.

**Bathroom Sink Use.** Weekday usage would be expected to occur at 630-730 and 2200-2300. The data show that bathroom sink use peaks at 600-800 (22%) and 2200-2400 (19%) at the upstairs sink and at 600-800 (23%) and 1600-1700 (13%) downstairs. Weekends find more activity at the upstairs sink in the late morning hours 800-1200 (49%) with the downstairs sink receiving only infrequent use mainly at lunch time (1200-1300, 21%). Bathroom sink use in this family patterns well and conforms visibly with their reported schedule.

**Shower Use.** Shower activity should occur mainly between 630 and 715 on weekdays with later morning showers during the weekend. The data confirm this indication, since 67% of showering events occur between 600 and 800 on weekdays and 67% of weekend showers occur between 700 and 1200. Showering activity is highly regulated by daily schedule in this household.

**Laundry Use.** The family did not report their schedule for doing laundry. Data indicate that laundry use is most likely to occur during weekends (67%). Laundry use during weekdays occurs mainly between 1900 and 2100 (42%). Laundry use during the weekends forms a near normal distribution around 1300-1400. Thus, laundry use, although to a lesser degree than showering, can be considered a patterned activity in this household.

### Family 5

Family 5 is a large household composed of Jonnelle (45), a nutritionist, her mother, Ora (68), and her children, Marcus (13), Nicole (12), Roschelle (8), and Alton (7). Jonnelle has a regular work schedule, and the children attend school. Ora is home all the time, usually in bed.

| Monday-Friday | | | |
|---|---|---|---|
| 545 | Jonnelle gets up, turns up thermostat | 1745-1800 | Jonnelle comes home, turns up thermostat |
| 600 | Jonnelle uses sink | 1815-1830 | Dinner preparation |
| 600 | Children get up | 1900-1930 | Dinner |
| 700 | Nicole showers | 2000 | Alton goes to bed |
| 715-720 | Jonnelle leaves, often turns | 2030-2115 | Nicole washes dishes |
| 730 | Marcus leaves, turns down thermostat | 2100 | Roschelle goes to bed |
| 800 | Nicole leaves | 2130 | Jonnelle showers |
| 820 | Alton and Roschelle leave | 2100-2400 | Marcus turns down thermostat |
| 820-1430 | House is empty except for Ora | | goes to bed |

**Thermostat Use.** This family reports that their thermostat doesn't control heating as well as they would like. They report that they feel cold and turn the thermostat up until the room becomes too warm, at which point they turn it down. They also report that there is temperature variation in different areas of the house. A window in one son's bedroom was broken and leaked air during part of the heating season. It is conceivable that in order to make different rooms comfortable, high settings are sometimes necessary. However, the house is well-insulated and the extreme settings we observed are hard to understand. The family insisted that they try to maintain a regular weekday schedule of nighttime and daytime setback. They report that they turn it up in the morning (545-600) and back down when Marcus leaves for school (730). It is then set up again when Jonnelle returns (1745-1800) until bedtime when Marcus sets it down (2000-2400).

Median hourly settings (Figure 8) show a weekday pattern of daytime thermostat setback (700-1500), and no evidence for nighttime setback. Weekends exhibit the same pattern, except that the daytime setback occurs between 1000 and 2000. This pattern is difficult to reconcile with their reported family schedule, however, since the thermostat is set down during times when only Ora is reported at home (she is said to never adjust the thermostat). A scatter plot of thermostat settings (Figure 9) indicates frequent thermostat adjustment and a very broad scatter around the median settings. Thus, there is no distinct pattern of thermostat setting in this house.

**Kitchen Sink Use.** Weekday kitchen sink use should be most prevalent between 1800 and 1900, and between 2000 and 2200.

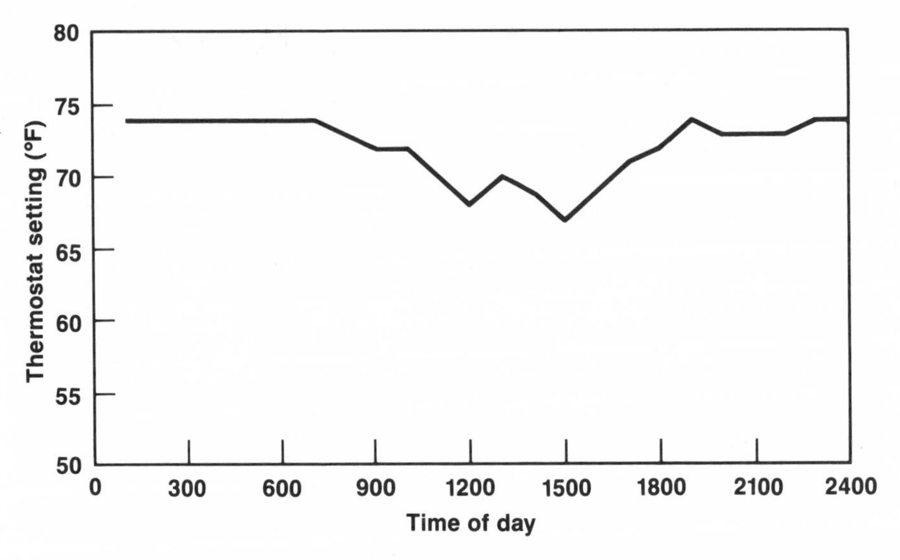


**Fig. 8. Median hourly thermostat settings, Family 5 (Monday through Friday).**

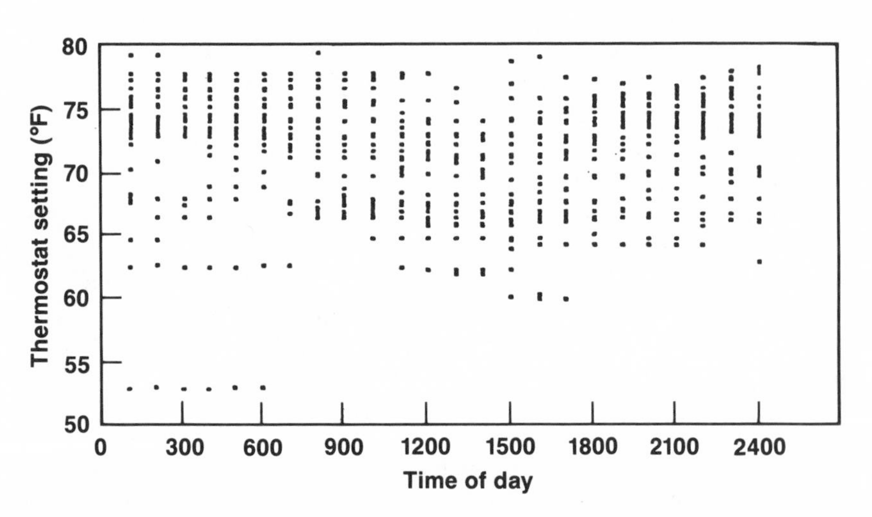


**Fig. 9. Thermostat setting scatterplot, Family 5 (Monday through Friday, hourly readings).**

The instrument data only partially confirm this conclusion, as sink uses actually take place most frequently between 700 and 1000 (26%), with a second peak occurring at 1800-2000 (23%). Kitchen sink use does exhibit patterns.

**Bathroom Sink Use.** Weekday bathroom sink use schedule was reported only by Jonnelle, who said she used it at 600-700. The downstairs sink shows heavy use at 600-800 (48%), and the upstairs sink shows a peak at 500-800 (41%). Thus, about half of all bathroom sink uses in this household occur in the early morning as the family gets ready to leave for the day.

**Shower/Tub Use.** Weekday showers reportedly occur at 700-800 and 2100-2200. Actually, only the morning peak is apparent since 48% of all showers take place at 600-800. Weekend showers are common and take place anytime during the late morning and early afternoon. As with the other families, shower behaviors are well patterned.

**Laundry Use.** No schedule for doing laundry was reported by the family. However instrument data indicate that this activity is patterned during weekdays, since 36% of all laundry activities take place between 700 and 900. Laundry activities during weekends are apt to occur at any hour.

## SUMMARY OF ANALYSIS

We can see that thermostat setting and hot water consumption behaviors are often patterned, and that the patterns are often related to the daily schedule of the occupants. It is evident that some behaviors are more clearly patterned than others, and that some patterns relate better to daily scheduling than do others. Whether particular behaviors are patterned by daily schedule in households depends on a number of factors, such as the type and complexity of daily schedules followed by family members.

According to the data, thermostat setting behavior is strongly influenced by when occupants are at home and the degree to which the schedules of the occupants coincide. This conclusion is particularly evidenced by Families 2 and 4, who have relatively fixed and simple schedules. Family members usually wake up and go to bed at about the same times, leave the house and come back at fixed times, and the houses are typically empty for regular periods of the day. Family 1, on the other hand, has a more erratic schedule since two members are almost always at home. Perhaps the most important behaviors that affect thermostat setting are getting up in the morning and going to bed at night.

Family characteristics appear to affect patterns of hot water use less than they do thermostat-setting behavior. The data show that many hot water use events are patterned, and that these patterns are often related to daily schedules. It appears that bathing/showering behavior is both most patterned and affected by scheduling. Even in households with complex family schedules, such as Family 1, this usage reflects the schedules of the occupants.

Kitchen sink use is also highly patterned and related to family scheduling. Only Family 1's kitchen sink use could be interpreted as being relatively unpatterned and non-reflective of the family schedule. Yet, if one considers frequent use throughout the day a "pattern", and Flora's scheduled activity from 800-1900 as "working in the kitchen", kitchen sink use has a definite relationship to the family schedule.

Bathroom sink use is probably the most erratic hot water use and has the least to do with daily schedule. Some households, however, such as Families 2 and 4, show a surprising degree of patterns. Laundry hot water use behavior is also surprisingly well patterned, as visible regularity was observed in 4 out of 5 cases.

It is not surprising that most families' schedules became increasingly flexible during the weekends and that this flexibility was reflected in their energy use behaviors. Visibly different patterns were found on weekends for many behaviors. Since people went to bed and woke up at later times, weekends typically shifted weekday patterns toward later hours.

## CONCLUSIONS

The behavioral component of residential energy consumption is an important topic for study. Thermostat setting and hot water consumption behaviors show a degree of patterning, and it appears that these patterns are related to the daily schedules of the occupants.

This chapter presents interview and instrumentation data to compare the reported schedules of four case study households with records of discrete behavioral events. The data indicate that several factors influence the patterns of these behaviors. The most prevalent factor is day of the week. For many behaviors, and across all families, weekends showed different types of energy use patterns than did weekdays.

For some families, energy behavior is highly patterned, and for other families only a few behaviors show patterns. With further research it should be possible to categorize families on the basis of a few variables and make reasonable guesses as to the degree of patterns in their energy behavior. We should expect that small families, those that leave and return to the house at regular and coincident times of day, and families whose houses are typically empty during the day will have highly patterned energy behaviors. Conversely, we would expect patterns to be less evident in larger families, families whose members have vastly different or irregular work schedules, and in families with members at home all day.

Since families have different schedules that affect how they consume energy, they may need to utilize different strategies for conserving energy. Families with consistent thermostat schedules should have conservation strategies available that utilize this consistency. Families with inconsistent or complex thermostat schedules could better save energy by utilizing strategies that are not affected by daily schedules. For example, Families 2 and 4 have regular schedules, and a clock thermostat would make thermostat setting as consistent as their daily schedules seem to be. Family 5, however, has an inconsistent and complex schedule. This household might find a clock thermostat useless, and simply defeat the clock mechanism by manual adjustment. A preliminary step might be to weatherize different areas of the house to reduce the temperature differences that might be causing inefficient use of the furnace through frequent thermostat adjustments.

As with thermostats, optimum hot water conservation strategies vary across households. Families with frequent use of one tap (like the kitchen sink in Family 1) might benefit greatly from insulating the pipes leading to that tap. Families with infrequent yet consistently scheduled water use might benefit from utilizing a clock device on the central water heater. Families having consistent, scheduled use at specific taps (like the shower in Family 2) might utilize time-of-day controlled point-of-use water heaters.

Since it should be possible to use several simple variables to categorize households on the basis of behavior patterns, it is feasible to use these variables to target the most effective behavioral conservation strategies to different households. RCS and other audit programs might utilize these variables to calculate the efficiency of different conservation strategies. Utilities could utilize these variables to target those households with electric heat and/or hot water that might best utilize off-peak utility rate schemes. The variables are probably also simple enough for consumers to use in self-help educational materials.

The influence of family schedules on energy-using behaviors deserves more research across a wider sample of households. Certain behaviors seem to be consistently patterned and such patterns precisely known by occupants; data regarding these behaviors might be successfully gathered using survey techniques. These behaviors include: setting the thermostat at night (Monday-Friday), setting the thermostat in the morning (Monday-Friday), showering/bathing, preparing and eating meals (Monday-Friday), and going to bed at night. Gathering schedules for other behaviors such as using bathroom sinks, doing laundry, setting the thermostat during the daytime and almost all weekend activities would probably be less successful using survey techniques and better accomplished through instrumentation.

**Acknowledgement**

*This work was supported by NSF grant BNS 82-10088, the MSU Agricultural Experiment Station as Project 3152, and the Kellogg Biological Station Small Farms Project at MSU. The Family Energy Project is housed at the Institute for Family and Child Study, College of Human Ecology, Michigan State University, East Lansing, MI 48824.*

**References**

Lipschutz, R.D., R. Diamond, and R.C. Sonderegger. "Some Technical and Behavioral Aspects of Energy Use in a High-Rise Apartment Building." In *Families and Energy: Coping With Uncertainty*, ed. B. M. Morrison and W. Kempton. East Lansing, MI: Institute for Family and Child Study, College of Human Ecology, Michigan State University, 1984.

Socolow, R. H. "Twin Rivers Program on Energy Conservation in Housing: Highlights and Conclusions." In *Saving Energy in the Home: Princeton's Experiments at Twin Rivers*, ed. R.H. Socolow. Cambridge, MA: Ballinger, 1978.

Weihl, J.S., W. Kempton, and D. DuPage. "An Instrument Package for Measuring Household Energy Management." In *Families and Energy: Coping With Uncertainty*, 1984 (see above).

Weihl, J.S., and W. Kempton. "Residential Hot Water Analysis: Instruments and Algorithms." *Energy and Buildings* 8:197-204, 1985.

# SECTION V:
# Interaction of Building Systems with Occupants

# SECTION V: Interaction of Building Systems with Occupants

*Willett Kempton*
*Princeton University*

## INTRODUCTION

The chapters in this final section analyze the interaction of occupants with the physical design of buildings and with their energy delivery and billing systems. Hackett studies the effects of a conversion from master metering to individual metering of electricity, while Diamond and Fagerson study the interaction of occupant behavior with building design, air infiltration, and heating systems. These rich analyses should prove rewarding both to those whose interests lie primarily in building systems and to those who have emphasized behavioral studies.

Diamond's impressive study combines data from energy billing, measurements of the structure, and interviews with the occupants. In a low-income housing project for the elderly, he found a ten-to-one variation in energy use between households. Design and materials were similar for these units, and construction quality varied little and did not correlate with energy use. The thermostat settings reported by the occupants also correlated poorly with measured energy consumption. Diamond explains that this poor correlation is due to factors such as reporting error, mechanical and design problems with the thermostat, poor understanding of

*Energy Efficiency: Perspectives on Individual Behavior*

the thermostat's operation by the occupants, and the use of supplementary heat sources. Some of these factors are particular to this community—for example, some of the elderly residents did not use windows for ventilation because they were too weak to open the latches, or because these individuals did not know how to operate the latches in these new homes. Diamond estimates that 40% of these households are using their gas stove for supplemental heat, in part because of inadequacies of the heating system.

Diamond selects three cases for further discussion, illustrating the measurable energy-consumption consequences of the management strategies reported by the occupants. He also divides the lowest and highest users from the entire sample, and compares their responses to selected questions. They were surprisingly accurate in their self-perceptions as high or low users. In response to questions about what they would do if too hot (summer) or too cold (winter), the low and high users reported the same frequency of using the air conditioning or heating. However, the low users more frequently reported alternative strategies, such as opening windows in summer or putting on more clothing in winter. This result suggests that lower energy consumption may be due more to a willingness and ability to practice alternatives, than to a reluctance to use the mechanical space-conditioning systems.

Hackett's chapter describes changes in energy consumption after the conversion of an apartment building from master metering to individual metering and billing. The average decrease in consumption was 36%, while a control group that did not undergo conversion increased consumption by 4%. Variation in the consumption decrease was analyzed across five social groups. All groups reduced consumption rapidly upon the billing change. The largest consumers—families with children and retired women—also showed the largest decrease in response to the billing change. Hackett uses interview data to interpret the residents' reactions to the change, and their conceptualization of energy use. While the billing change could have been seen as "freeing the frugal majority from the burdens of the self-indulgent few", interviews showed that it was regarded as benefitting the local utility or the apartment management.

In the closing sections Hackett suggests several intriguing possibilities that merit further thought. He raises questions about the social factors that motivate conservation, how home energy management may be seen as a reflection on the home dweller's reputation, and the effects of the current U.S. shift toward single-resident households and non-familial housing.

Fagerson examines data on physical construction, air infiltration, energy records, and questionnaire responses to sort out the interaction of housing design, occupant behavior, and energy use. Fagerson finds significant differences in energy use among nominally identical houses. Like Diamond, Fagerson finds that energy differences do not correlate well with infiltration as measured by pressurization tests in nominally identical houses. Fagerson then turns to the management or lifestyle variables, and finds her sample divided into two groups. In houses whose energy consumption is strongly affected by heat loss (e.g., passive solar with high amounts of glazing), consumption will be more strongly affected by thermostat setting, while consumption in well-insulated houses with smaller glazing areas is more determined by the number of hours of occupancy (due to heat gains from both occupants and appliance use). Although Fagerson had only self-reported thermostat figures available, she found that thermostat setting was the best predictor of energy use within groups of identical houses. Finally, Fagerson notes that building design may also affect management practice. In particular, houses with less insulation or improperly installed insulation may have colder wall surfaces, thus requiring higher interior temperatures in order to maintain the same level of perceived comfort.

# Energy Use Among the Low-Income Elderly: A Closer Look

*Richard C. Diamond*
*Lawrence Berkeley Laboratory*

## INTRODUCTION

As part of our continuing study of energy use in public housing, we have been studying how occupant behavior effects energy consumption. We initiated this study in 1982, specifically to find out how energy use is affected by actions taken during the design, construction, and occupancy of a building (Diamond 1983). The study site is a 60-unit public housing project for the elderly called Winston Gardens, which is located in Oroville, California. The sixty units at Winston Gardens, which were completed and occupied in August 1982, are similar in size (58 $m^2$ or 625 $ft^2$), number of residents per unit (three couples, the rest all single), and type and number of appliances. Thirty-five of the units have been continuously occupied by their original residents for eighteen months; these units constitute the main sample for most of the findings in this study. Each apartment has a gas water heater and a gas stove, a refrigerator, and an electric heat pump for heating and cooling. The residents are mostly single women (75%) between the ages of 55 and 85 whose individual monthly income is less than $300.

We selected this particular housing project for three reasons: its location (the area has an extended heating season and an extended

*Energy Efficiency: Perspectives on Individual Behavior*

cooling season), our previous collaboration with the project's architect (who provided us with considerable information on its design and construction), and its resident population of low-income elderly (selected because they are particularly vulnerable to rising energy prices and because little is known about the energy and comfort needs of this group). Over the past year we have amassed considerable data, both factual and anecdotal, on weather conditions, structural building parameters, computer simulations, occupant behavior, and utility bills. Our initial examination of the monthly, individually metered utility bills showed a large variation in energy use among what we had assumed were identical apartments with similar residents. The variations in energy use among the residents, in summer as well as in winter, are on the order of ten-to-one. To account for these large variations, we examined two standard physical determinants of energy usage: climate and construction quality.

## PHYSICAL DETERMINANTS OF ENERGY USE

Standard engineering models for determining energy consumption are based on two factors, a measure of the severity of the climate (given by the number of degree-days) and a physical index for the thermal integrity of the building. Our earlier work (Lipschutz 1983) presented evidence that the more severe the climate, the less variation in energy use among similar houses can be expected. Given that Oroville has a relatively mild climate—1,600 heating degree-days °C (2800°F) and 890 cooling degree-days °C (1600°F)—we would expect that there could be a large variation in individual energy consumption (which we observed), even though degree-days were an accurate indicator of aggregate energy consumption (see Figure 1). Figure 1 shows the correlation between the average monthly electricity consumption and heating degree-days for the 35 continuously occupied units at Winston Gardens. Electricity consumption figures recorded here include energy used for space heating (the electric heat pump) and baseload energy use (lighting and appliances). Because all units have identical appliances and are occupied by single residents whose lifestyles are similar, we assume that the baseload does not vary much from unit to unit.

The second variable in the standard engineering model is the thermal integrity of the building shell, which is made up of two parts, the thermal characteristics of the building components and

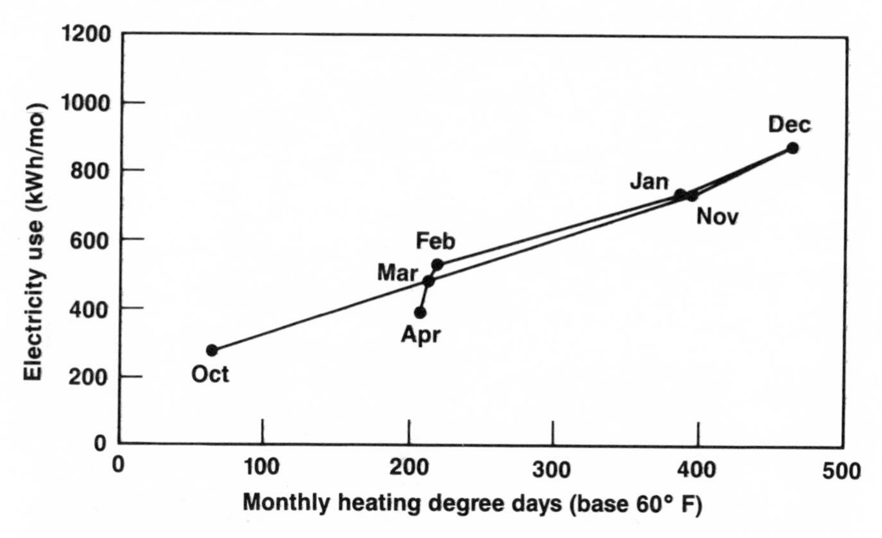


**Fig. 1. Heating season degree days (1982–83) and average electricity use for 35 units at Winston Gardens, Oroville, California.**

the tightness of construction. Because all the units at Winston Gardens are built using similar materials, we chose to examine the tightness of the individual units to look for hidden variations in construction quality. All but three of the sixty units were pressurized and depressurized using a blower door to move air through the units. By measuring the air flows and pressures during the tests we could then calculate the total leakage area for each unit. Our hypothesis was that leakier units—having large leakage areas—would have greater energy consumption in both summer and winter, given that residents would have to run their heat pumps longer to heat and cool the rooms.

The distribution of leakage areas showed that the units, although quite leaky (mean = 566 $cm^2$), were uniformly leaky (standard deviation = 98 $cm^2$ or 17%), and thus tightness of construction could not account for the variation in electricity use during the heating season. Furthermore, the correlation between leakage areas and electricity consumption was near zero (see Figure 2).

## BEHAVIOR VARIABLES

### Thermostat Settings

With the standard engineering indicators of building and climate

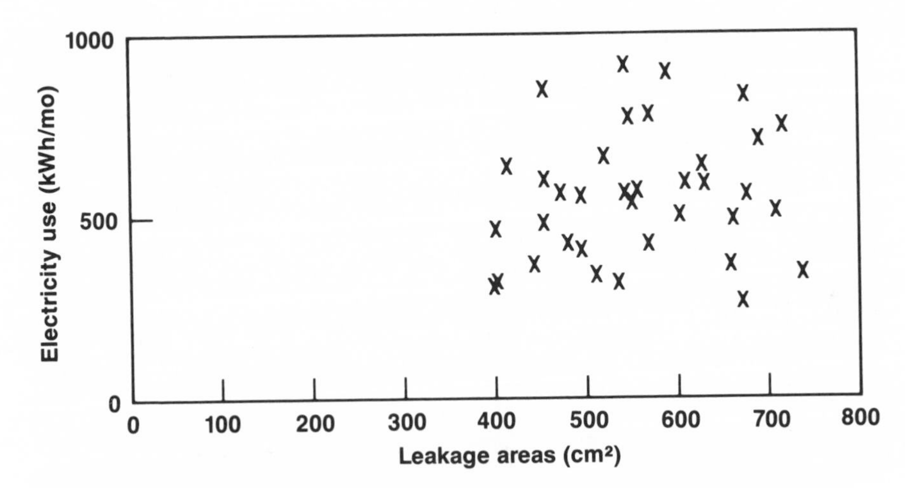


**Fig. 2. Leakage areas and average electricity use over the heating season (1982–83) for units at Winston Gardens.**

parameters unable to explain the variation in energy consumption at Winston Gardens, we tested thermostat settings as a potential predictor for energy consumption. Occupant behavior regarding thermostat operation is rather complex (Kempton 1983), and we had only limited data from our interviews on reported settings and usage patterns. (In addition to the survey data we had single readings of the summer air-conditioning settings that were observed during the house air leakage tests.) Figures 3 and 4 show heating season (3) and cooling season (4) electricity use plotted against thermostat settings (occupant-reported for winter, and observed for summer).

The low correlations between thermostat settings and energy used for space conditioning could be due to a number of factors: the unreliability of occupant-reported settings (through either ignorance or the possible inclination of some occupants to give the impression of correct behavior), improper calibration or poor design of the thermostat, and the occupants' poor understanding of thermostat operation, usage patterns in which the temperature setting is left unchanged but the thermostat is turned off for extended periods of time, and the use of supplementary heat sources, such as the kitchen stove. All of these situations were observed at Winston Gardens and may well explain the poor correlation the data show between the thermostat settings and electricity consumption.

A particular problem with the thermostats at Winston Gardens is

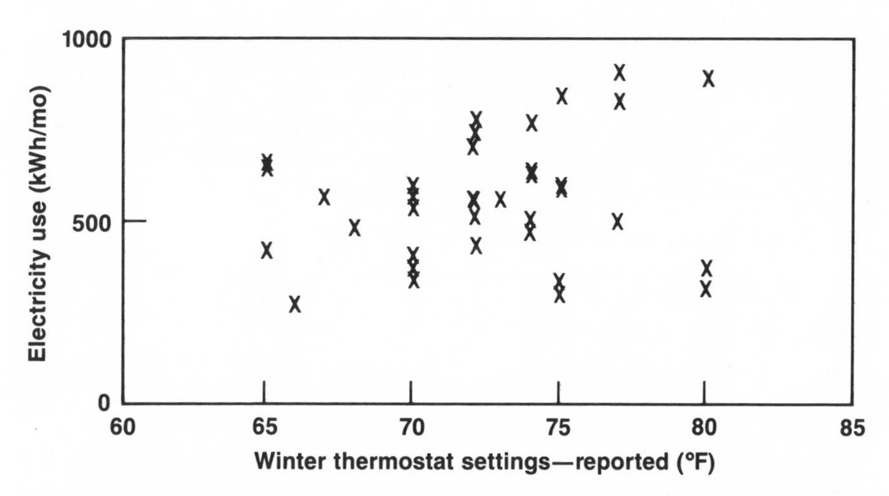


**Fig. 3. Winter thermostat settings (reported) and average electricity use, November 1982–April 1983, Winston Gardens.**

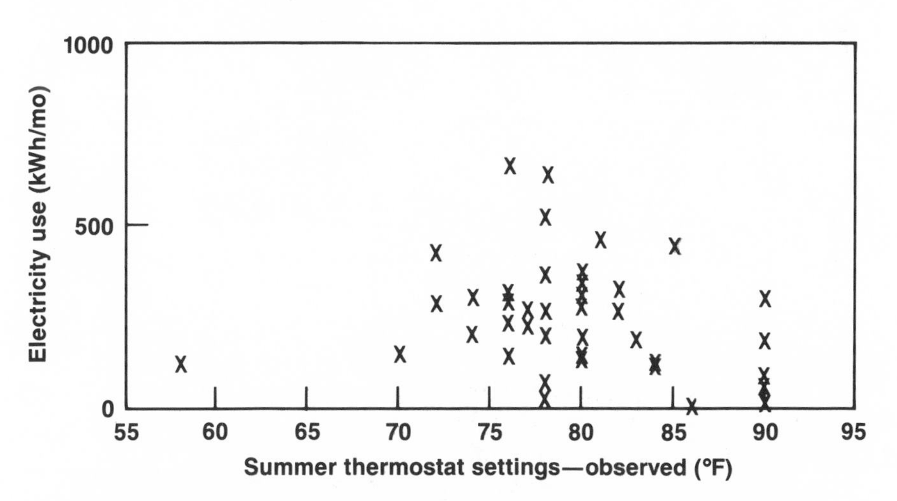


**Fig. 4. Summer thermostat settings (observed) and average space cooling electricity use, July 1983–September 1983, Winston Gardens.**

that the heat pump can be overridden easily by the resident turning on the emergency back-up electric-resistance heaters. This behavior is a result of either accidental usage, where the resident incorrectly sets the thermostat, or intentional design, by residents who prefer the heater strips to the heat pump because the air temperature is considerably higher, allowing the apartments to heat more quickly.

## Patterns of Energy Use

Having eliminated construction quality and thermostat settings as explanatory variables, we looked more closely at patterns of individual energy use, taking the average usage for all the units and examining individual deviations from that average. Figure 5 shows the average monthly electricity use for the 35 continuously occupied units.

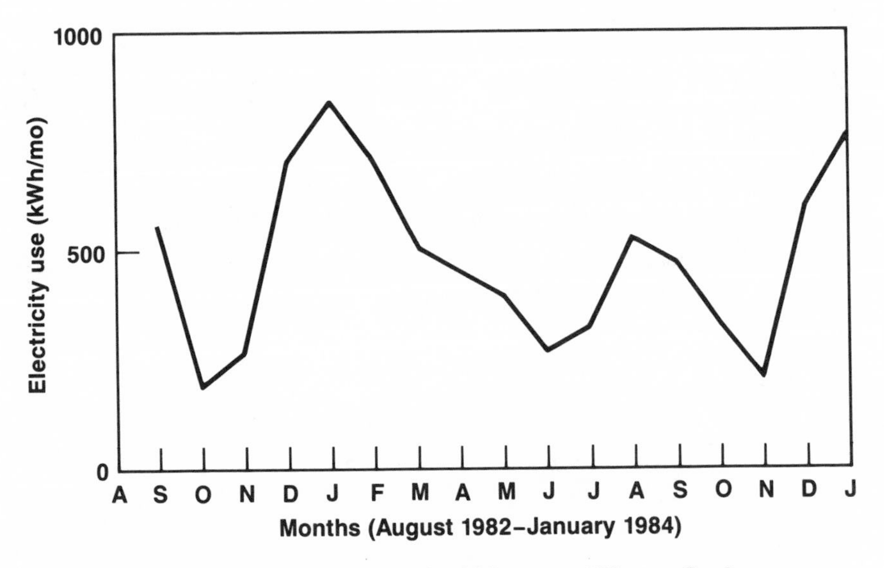


**Fig. 5. Average monthly electricity use for 35 houses at Winston Gardens.**

The data show two peaks: the first, between November and May, represents the electricity used for space heating; the second, between June and October, represents energy used for space cooling. A baseload calculation of 220 kWh/month for the refrigerator, lights, and television corresponds to the lowest dips in the graph. When individual units were plotted (Figure 6), the large variation among the identical users appeared dramatically.

Because the interview data showed little variation in the cooking and bathing practices among the residents, we expected a relatively constant yearly gas usage, and were consequently surprised by the seasonal variation in the average monthly gas consumption (see Figure 7).

The baseload value calculated for bathing and cooking, 8 therms/month, corresponds well with actual usage over the five summer months. Unexpected, however, was the finding that gas

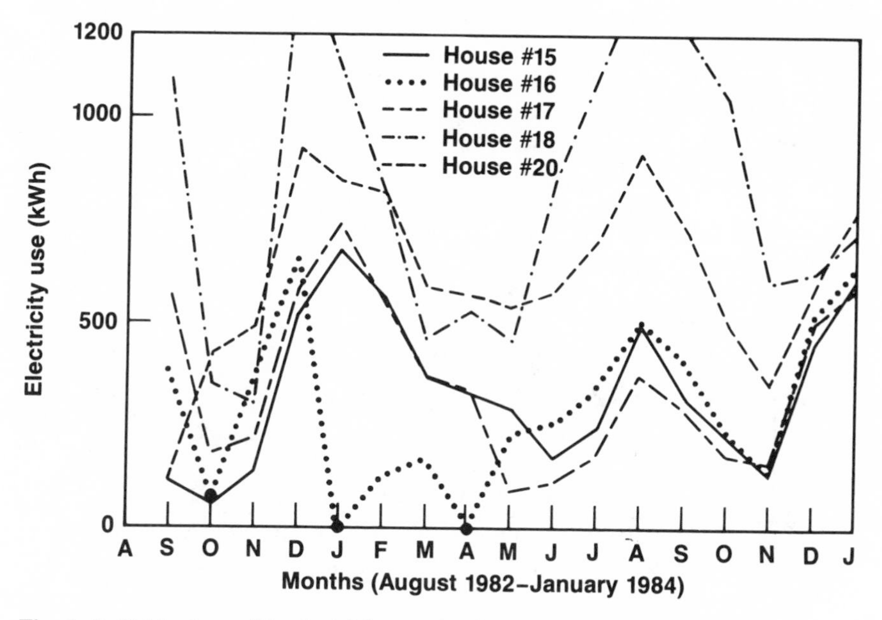


**Fig. 6.** Individual monthly electricity use for five units at Winston Gardens.

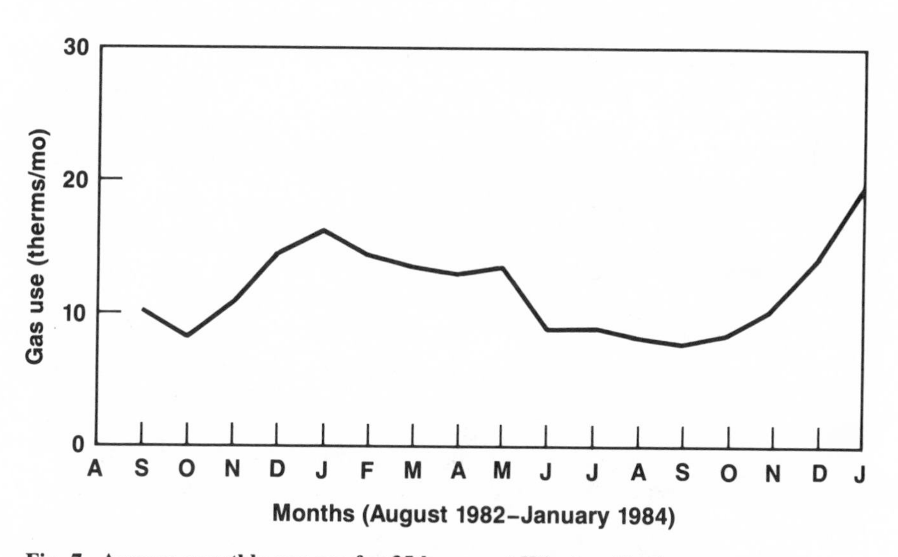


**Fig. 7.** Average monthly gas use for 35 houses at Winston Gardens.

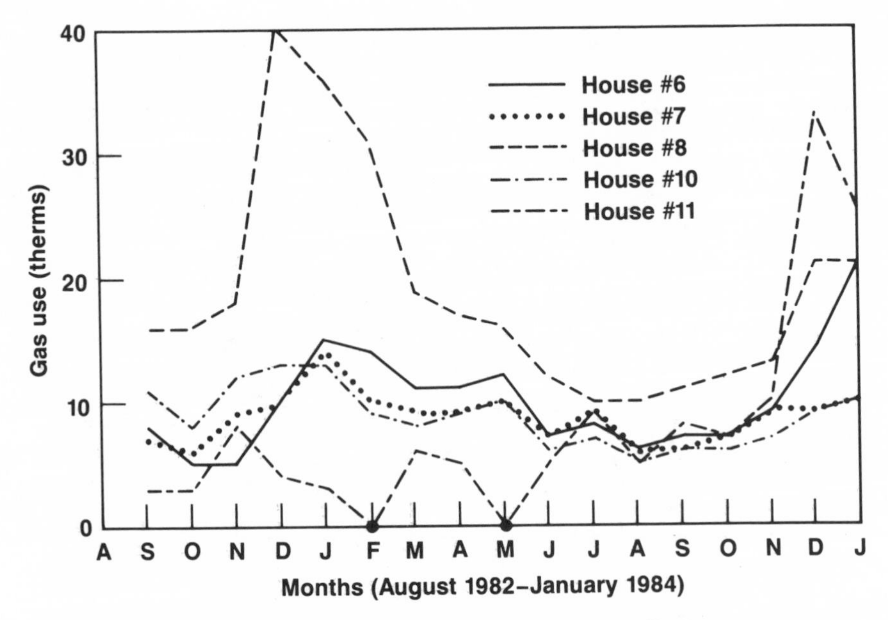


**Fig. 8. Individual monthly gas use for five units at Winston Gardens.**

usage increased 100% over summer usage for both winter periods. Possible explanations are: increased use of gas for cooking over the winter months; longer showers and baths in winter; colder incoming water at the water heater in wintertime, increasing gas consumption at the water heater; and the use of gas stoves for space heating. Plots of the individual units (Figure 8) support the last hypothesis, and interviews with the residents confirm it.

### Patterns of Individual Use: Case Studies

Following are case studies of four Winston Gardens residents based on interviews conducted with all the residents in August 1983. The interviews were based on a one-hundred-item questionnaire that addressed several aspects of the residents' living patterns and housing. Accompanying each case study is a figure with eighteen months of electricity consumption plotted as a ratio of the individual's usage to the group average. The cases were selected to represent different categories: higher than average, lower than average, low winter/high summer, and a low user with gas stove used for additional heating.

- *Case 1.* (Figure 9): An 87-year-old widow who had moved to Winston Gardens from an apartment in another part of the county.

She left because of the "poor ventilation in the room and rowdiness of the neighbors."

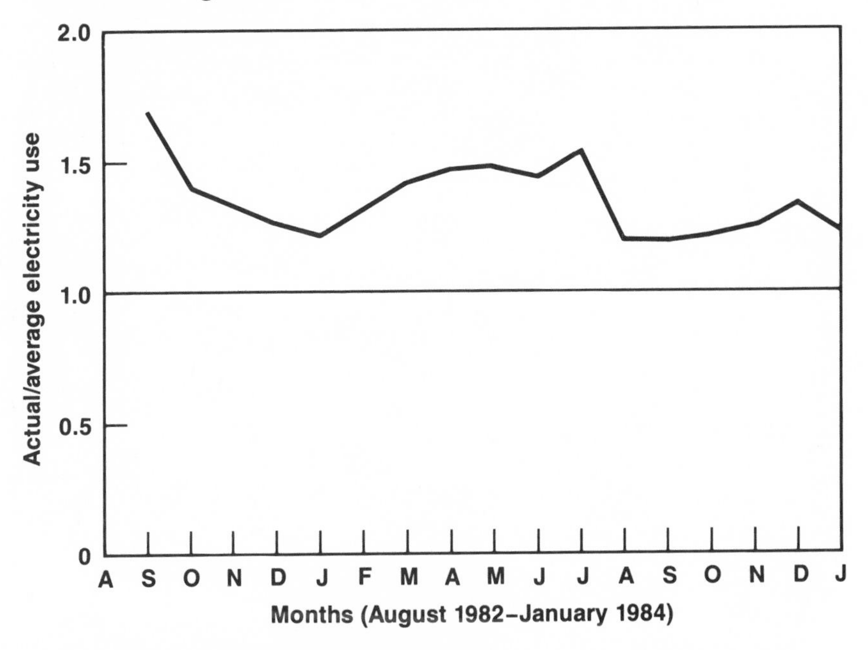


**Fig. 9. Case 1, actual/average electricity use.**

She keeps the thermostat set at 23 °C (74 °F) all winter and never moves it ("I have to have it that warm"). She has carpeted over the hall and kitchen floor because the tile on the concrete slab floor was too cold. In summer she uses the air conditioner: "I can't open the windows, I'm too crippled." Unlike most other residents, she leaves the drapes open during the day in summer because she likes to look at her flowers. Her television is on every day from 5:00 a.m. to midnight. Excluding the heat pump, the television is the single largest user of electricity (2.3 kWh/day), consuming more than the refrigerator (2.0 kWh/day). She attributes her high energy usage—36% higher than the average at Winston Gardens—to the air conditioner, which she claims to use more than the others, and to her toaster oven and microwave.

• *Case 2.* (Figure 10): A 71-year-old, who had moved to Winston Gardens from her apartment in town when the rent was raised. Unlike Case 1, she uses 42% less energy than the group average. In winter she keeps the house at 19 °C (66 °F) during the day, and shuts the heat off at night. "My friends say, 'My, aren't you freezing to death?' and I say, 'No, I'm comfortable." In summer she keeps drapes closed and windows open all the time; "I hardly ever use the air conditioner." She also has rigged up a blind on her west-facing porch to provide shade from the hot afternoon sun. The baseload usage (as indicated by the two peaks occurring in the fall) is 10% less than average.

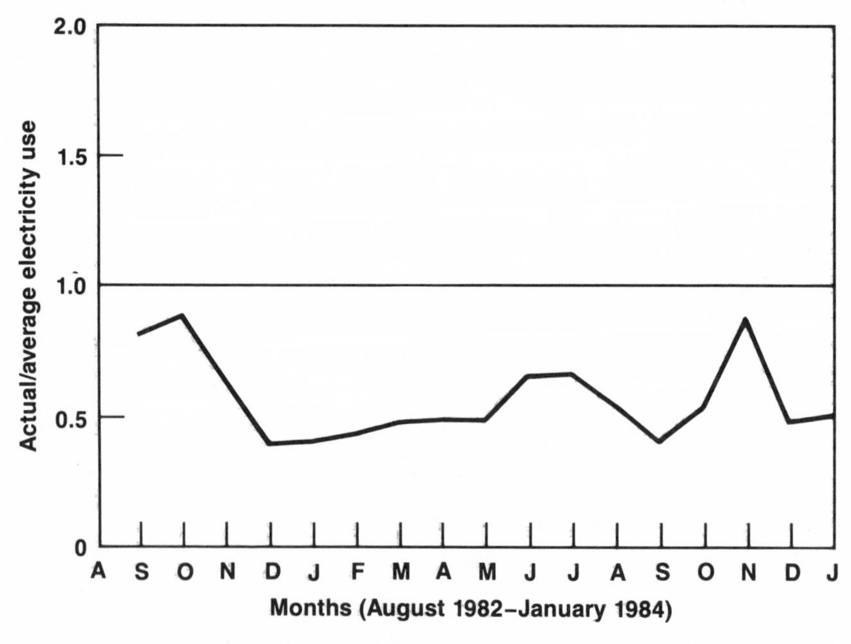


**Fig. 10. Case 2, actual/average electricity use.**

• *Case 3.* (Figure 11): A 74-year-old woman, she is also a low consumer of electricity—17% less than the average.

Her winter usage is right on average, but her summer use is substantially lower than average; as she says, "I never use the air conditioner. I can't stand that thing blowing down on me." In summer she keeps the windows open as well as the drapes because "I like the view." While the electricity data here suggest a low energy user, the gas data tell a different story: her gas consumption in winter is three times the average for the group. Because she says she has no change in bathing or cooking practice throughout the year, we assume the gas use is from the gas stove used for space heating. She admits that, although she does no baking, she dislikes the stove because "you can't light a burner when the oven is on."

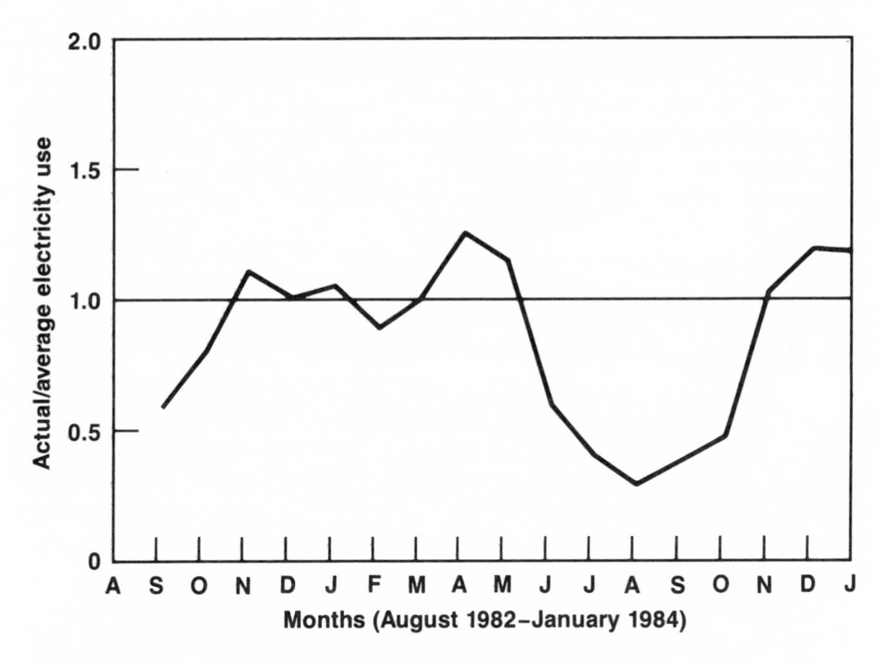


**Fig. 11. Case 3, actual/average electricity use.**

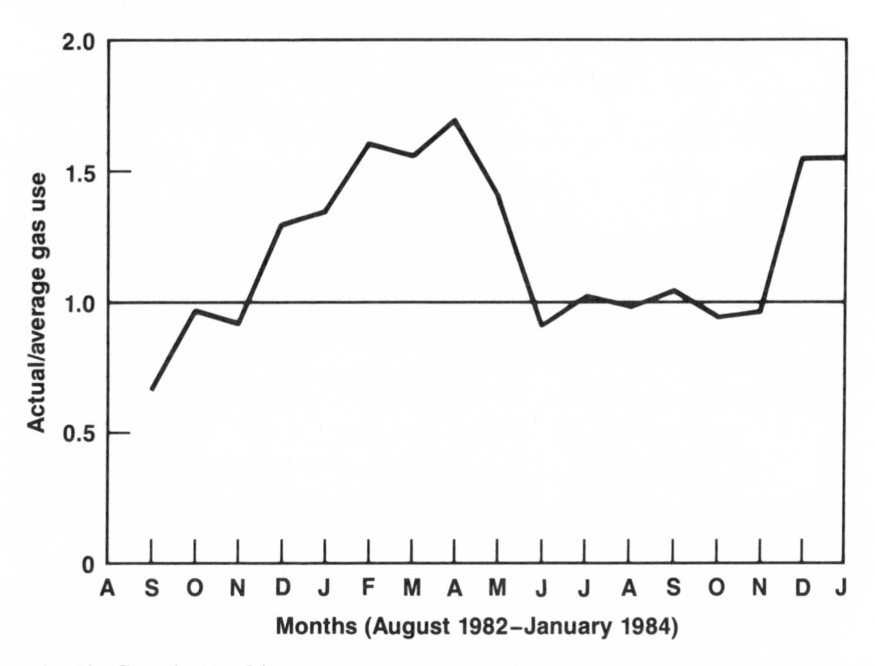


**Fig. 12. Case 4, actual/average *gas* use.**

• *Case 4.* (Figure 12): A 72-year-old man who says he turns on the kitchen burners for warmth. The elevated gas use in the winter season is clearly seen in the figure.

As he explains, the electric heating is too noisy and expensive. He adds: "I don't need as much [heat] as those women." In summer he uses four floor fans, opens windows, and closes the drapes, using the air conditioner only in the evenings. His electricity use (not shown) is 45% less than the average.

**Patterns of Individual Use: High Users vs. Low Users**

To look for overall behavioral patterns, we split the plots of the actual/average data into two groups: high users and low users. We then looked at characteristics of each group for explanatory indicators. The characteristics we identified are summarized in Table 1. Both high and low users were fairly accurate in guessing their relative usage, even though there is little socializing among the residents, and as one woman put it, "We never talk about things like that."

Asked to give reasons why their usage differed from those of their neighbors, high users cited health reasons, frequent use of the air conditioner, and not caring about conserving. Low users usually said they did not need as much air conditioning and that, generally, they were more conservative than the high users. The responses of these two groups to the questions, "What do you do when you are too hot?" and "What do you do when you are too cold?" are given in Table 2.

Differences between the two groups are that low users say they are more likely to wear additional clothes in winter and to open windows, close drapes, and use floor fans in summer rather than turn on the air conditioning. Several women in the high-user group mentioned not being able to open the windows because of arthritis as the reason for using the air conditioning. High users gave as reasons for not pulling drapes in summer: liking the natural light, preferring to be able to see out, and not knowing how to operate the blinds.

**Explanatory variables.** From the interviews with the residents of Winston Gardens, we have identified five categories of variables to explain the variation in energy consumption. The categories are: 1) the residents' health and comfort, 2) the level of satisfaction with the heating/cooling system, 3) the degree of understanding and control exercised, 4) attitudes toward conserving, and 5) income and status.

**Table 1. Characteristics of low and high energy users at Winston Gardens.**

| | Low Users (n=22) | High Users (n=13) |
|---|---|---|
| **Sex** | | |
| male | 31% | 27% |
| female | 69% | 73% |
| **Age** | | |
| mean | 71.4 years | 68.3 years |
| st. deviation | 7.2 years | 8.7 years |
| **Perceived energy use compared to neighbors** | | |
| more | 5% | 46% |
| about the same | 27% | 15% |
| less | 59% | 0% |
| not sure | 9% | 39% |
| **Use gas stove for heating** | 45% | 25% |

**Table 2. Behavioral responses for heating and cooling for high and low users at Winston Gardens.**

| | Low Users (n=22) | High Users (n=13) |
|---|---|---|
| **Heating** | | |
| use furnace | 68% | 69% |
| wear more clothes | 45% | 23% |
| use space heaters | 5% | 23% |
| **Cooling** | | |
| use air conditioning | 73% | 69% |
| open windows | 91% | 31% |
| close drapes | 50% | 30% |
| use floor fans | 32% | 15% |

**Health and comfort.** Part of our initial impetus for studying this particular population was to find out something about the thermally related health and comfort of the elderly. Only four of the residents said they required heating or air conditioning for their health. Some residents who said they did not mind the heat or that they were never cold would comment on their neighbors' wasteful behavior. The wide range of interior temperatures recorded during the summer interviews, 22 to 35 °C (72 to 95 °F), may be more indicative of relative income, however, than of temperature preference and comfort.

Fanger has shown in his studies of thermal comfort that there are no general differences in temperature preference based on age or sex, and that an individual's comfort in a given environment is determined by her activity level and type of clothing (Fanger 1970). According to residents, daily activities at Winston Gardens are chiefly reading, sewing, and watching television, all sedentary activities. Typically the women wear thin cotton dresses with either a light sweater or a robe in winter. Under these conditions their temperature preferences in winter should be higher than the reported mean winter thermostat setting of 22 °C (71 °F). One way to increase thermal comfort is to use supplementary space heaters, and we believe that the residents' high use of these heaters is linked to their dissatisfaction with their heat pumps.

**Satisfaction/dissatisfaction with the heating/cooling system.** Only 55% of the residents say they are satisfied with the heating system, and 63% with the cooling system. The problems cited are the noise, the drafty air, the cost, and the system's slowness in heating up. A British survey on the heating preferences of the elderly concluded that radiators and floor systems were the most preferred and ducted warm-air systems the least—too drafty and too noisy, and the controls were considered difficult to understand (Page and Muir 1971). In addition to these factors is the particular characteristic of heat pumps that the warm air coming out of the registers, because it is only a few degrees warmer than room temperature, feels cool. As a local, radiant, familiar and easily controlled heat source, and one that is considerably cheaper to operate than the electric heat pump, the gas stove's use for space heating is not surprising. We estimate 40% of the residents use their stove for some space heating. Because the use of these stoves for space heating represents a potential health and safety hazard, we have initiated a pilot retrofit project with the Housing Authority to modify the heat pumps and bring heat to the kitchen directly.

One unusual and disturbing finding at Winston Gardens has

been the high number of residents (30%) who have moved away in the first year. Several reasons were given by the manager and neighbors for the cause of their leaving, chief among these being dissatisfaction with the heating and cooling system. Previous studies have shown that residents of public housing often singled out mechanical systems for blame, not only because of real deficiencies in their operation but also because they are easily identifiable objects to criticize and condemn (Cranz 1977). According to other studies, residents who feel stigmatized by living in public housing and whose general life satisfaction is low will target mechanical systems rather than criticize management or their own situation (Becker 1977). Actually, at Winston Gardens, residents rate their life satisfaction as fairly high, and, with few exceptions, think well of the management. Because there are valid criticisms of the mechanical performance of the heating/cooling system at Winston Gardens, we have not determined to what degree other confounding variables might explain the high user dissatisfaction. Contacting residents who have moved out may prove useful in this regard.

**Understanding and control.** One factor that emerged from the interviews was the degree to which the residents abandoned control over their situation. Several mentioned never touching the thermostat: "My son-in-law sets it for me" or "I get the manager to move it." In many cases, these same residents never moved or adjusted their mini-venetian blinds: "I don't know how to adjust them;" "I never touch anything I don't understand"; or "I just leave them the way they are, I like 'em just fine." One woman said, "There's nothing wrong with the thermostat, only the dummy who tries to use it."

Several of the residents expressed dissatisfaction with their showers, especially the hand-held shower head and the inability to control the water flow. One woman had been taking baths simply because she did not know how to work the shower, which only required a knob to be pulled up. In most of these cases, tenants were willing to do something–if only they knew what to do; afraid of doing the wrong thing, many simply lived with the problem.

**Attitudes toward conserving.** The elderly, much like the rest of the population, have mixed feelings about energy conservation. Often these sentiments differ with age and income (Minnesota Department of Energy and Economic Development). Younger seniors (ages 60-70) are more aware of conservation and more receptive to conservation messages. Like many of their younger counterparts, seniors with sufficient income to meet their living expenses often do not want to be bothered with conservation actions. One fairly well-off respondent reported during the interview, "I don't mind P.G. & E. telling me to save energy, they can do what they want. I just ignore them." The main concerns voiced by residents were: general uncertainty about technology, personal health, financial solvency, and the length of time they will be able to live independently.

Many residents believe they are doing everything they can to keep energy costs down, to the point where they are miserably cold in winter and uncomfortably hot in summer. The goal of energy conservation work here should be to ensure that comfortable temperatures are maintained while energy is saved. Residents pay their utility bills directly to the utility, with the Housing Authority providing a minimum utility allowance as a reduction in the rent. Another factor in the residents' attitudes toward conserving is the difference in the summer and winter allowances for the cheaper "life-line" rates. One reason residents think they use more energy for cooling than for heating is that they have a smaller quota for summer life-line usage, and consequently end up paying more.

**Income and Status.** Income has been mentioned previously as a possible factor underlying differences in temperature preferences. Being able to heat and air-condition the house for long periods of time has become a status indicator for many residents. Residents can hear the frequency and duration of their neighbors' heat-pump operation, and, as mentioned previously, commented on the high and low users during the interviews. Knowing that all residents must be below a certain income level to be eligible for Winston Gardens, we did not think that income would constitute an important variable in this research. However, residents are allowed to have up to sixteen thousand dollars in savings in addition to their income, and it is possible that differences in financial security are behind the behavioral variations we noted.

## CONCLUSIONS

Our study of energy use among the low-income elderly at Winston Gardens continues to perplex and astonish us. We have found that simple indicators such as reported thermostat settings and measured air leakage area are not able to explain the variations in energy use observed in the utility bills. By examining the energy use of individual residents and of categories of high and low energy users, we have identified characteristics of each group that help explain their different usage patterns. Differences in the residents' health and comfort, their level of satisfaction with the heating/cooling system, the degree of understanding and control they exercised, their attitudes toward conserving, and their income and status all appear to be variables underlying the wide divergence in energy usage. A next step will be to use these variables in developing a model that can characterize behavioral effects on energy consumption. Given our experience with the data from the residents at Winston Gardens, this will be no small task.

**Acknowledgements**

*The author wishes to thank Peter Cleary, Helmut Feustel, Chuck Goldman, Willett Kempton, and Max Neiman for their comments in reviewing this paper; Laurel Cook for her editorial work on it; the Center for Environmental Design Research for administering the project; and the architects, managers, and residents of the housing projects for their cooperation.*

**References**

Becker, F.D. *Housing Messages.* New York: Dowden, Hutchinson & Ross, 1977.

Byerts, T.O., S.C. Howell, and L.O. Pastalan, eds. *Environmental Context of Aging: Life-styles, Environmental Quality, and Living Arrangements.* New York: Garland STPM Press, 1979.

Claxton, J.D., C.D. Anderson, J.R. Ritchie, and G.H. McDougall, eds. *Consumers and Energy Conservation.* New York: Praeger, 1981.

Cranz, G., D. Christensen, and S. Dyer. *San Francisco's Public Housing for the Elderly.* Center for Environmental Design Research, University of California, Berkeley, 1977.

Diamond, R.C. "Energy and Housing for the Elderly." Paper presented at the Conference on Families and Energy, Michigan State University, October 1983.

Fanger, P.O. *Thermal Comfort.* New York: McGraw Hill, 1970.

Grier, E. "Energy Pricing Policies and the Poor." In *Energy and Equity,*, ed. Ellis Cose. Washington, DC: Joint Center for Political Studies, 1979.

Hedlin, C.P. and Bantelle. "A Study of the Use of Natural Gas and Electricity in Saskatchewan Homes." In *Proceedings of the Technical Program of the 91st Annual EIC Meeting*, Jasper, Alberta, NRC Report 16898, May 1977.

Kempton, W. "Two Theories Used for Home Heat Control." Paper presented at the Conference on Families and Energy, Michigan State University, October 1983.

Kempton, W. and L. Montgomery. "Folk Quantification of Energy." *Energy* 7:817-827 1983.

Lipschutz, R.D. and R.C. Diamond. "Energy Use in a High-rise Apartment Building." Lawrence Berkeley Laboratory, LBL-16366, September 1983.

Minnesota Department of Energy and Economic Development. "Energy Information for the Older Population: A Market Research Report." Available from the Energy Division, 980 American Center Building, 150 East Kellogg Blvd., St. Paul, MN 55101.

Page, D. and T. Muir. "New Housing for the Elderly." London: Bedford Square Press, 1971.

Seligman, C., J.M. Darley, and L.J. Becker. "Behavioral Approaches to Residential Energy Conservation." *Energy and Buildings,* 1:325-338 1978.

Socolow, R., ed. *Saving Energy in the Home: Princeton's Experiments at Twin Rivers.* Cambridge, MA: Ballinger, 1978.

Sonderegger, R.C. "Movers and Stayers: The Resident's Contribution to Variation Across Houses in Energy Consumption for Space Heating." *Energy and Buildings,* 1:313-324, 1978.

Struyk, R.J. and B.J. Soldo. *Improving the Elderly's Housing: A Key to Preserving the Nation's Housing Stock and Neighborhoods.* Cambridge, MA: Ballinger, 1980.

U.S. Department of Housing and Urban Development. *Low Rise Housing for Older People, Behavioral Criteria for Design.* Prepared by Zeisel Research. Washington, DC: U.S. Government Printing Office, 1977.

Warriner, G.K. "Electricity Consumption by the Elderly," In *Consumers and Energy Conservation*, ed. J.D. Claxton et al. ed. New York: Praeger, 1981.

Wilk, R. and W. Wilhite. "Household Energy Decision Making in Santa Cruz County, California." Universitywide Energy Research Group Report, UER-107, University of California, October 1983.

Winett, R. et al. "A Field-based Approach to the Development of Comfort Standards, Energy Conservation Strategies, and Media-based Motivational Strategies." *ASHRAE Transactions,* vol 89, 1983.

# Energy Consumption and Energy Billing in Apartments

*Bruce Hackett*
*University of California, Davis*

## INTRODUCTION

This chapter reports the most recent findings of a continuing study of electrical energy consumption in apartments in Davis, California. The research interprets the drop in consumption following a change from master to individual metering and billing of electricity, and investigates the relationship between energy use and both the structural and social features of apartments.

In January 1983, the Anderson Place apartment complex switched from master to individual apartment billing of electricity—the apartments are all-electric, with all cooling and most of the heating supplied by a single living-room wall-mounted unit—and this event was followed by a 36% drop in kilowatt-hour consumption in 1983 compared with the previous year. The drop was, in fact, substantially greater than 36% if we compare it with the consumption of three "control" apartment complexes in Davis. Table 1 shows this comparison for each of the first fourteen months following the change. This altered consumption pattern is generally in line with that predicted by studies elsewhere (Nelson 1981; McClelland 1983), but the magnitude of the drop may reflect in part the peculiarities of this setting. These are relatively low-

*Energy Efficiency: Perspectives on Individual Behavior*

**Table 1. 1983-1984 kWh/Day/Unit Compared to 1982 Consumption by Month for Anderson Place and Controls (Percentage Change)**

| Month | Anderson | Average Of Three Controls | Differences |
|---|---|---|---|
| Jan 1983 | -28 | 5 | 33 |
| Feb 1983 | -44 | -4 | 40 |
| Mar 1983 | -36 | 0 | 36 |
| Apr 1983 | -45 | 6 | 51 |
| May 1983 | -26 | 10 | 36 |
| Jun 1983 | -29 | 17 | 46 |
| Jul 1983 | -37 | 5 | 42 |
| Aug 1983 | -32 | 12 | 44 |
| Sep 1983 | -33 | 4 | 37 |
| Oct 1983 | -28 | 21 | 49 |
| Nov 1983 | -44 | -4 | 40 |
| Dec 1983 | -33 | -4 | 29 |
| Jan 1984 | -43 | 0 | 43 |
| Feb 1984 | -44 | -8 | 36 |
| Average | -36 | 4 | 40 |

income apartments (FHA Section 236 provides a *de facto* rent subsidy that at present amounts to $53 a year on a one-bedroom unit); Davis has a reputation as an especially energy-conscious community; and the town's temperate climate, with a relatively small indoor-outdoor temperature differential throughout the year, means that even a small shift in indoor temperature can cause a fairly large change in consumption.

The complex harbors a mix of what we came to call "life-cycle" groups, characterized as follows:

1. one-resident apartments occupied by one younger man (N=64),
2. one-resident apartments occupied by one younger woman (N=44),
3. one-resident apartments occupied by one older, retired woman (N=29),
4. apartments with more than one adult, but no children (N=52), and
5. families with children (N=50).

At the time that we gathered data, there were no older, retired men in the complex living in single-resident apartments.

Sorting the population in this way almost perfectly controls for apartment size—one bedroom in the first three categories, two bedrooms for the latter two. We were interested, then, in the different experiences of these groups in adapting to the billing change, and in general we found that all of them responded significantly to it—that is, adjusted their energy use downward. Figure 1 traces the consumption per apartment for each life-cycle group in the months subsequent to the January 1983 change. The figure shows an impressive drop from January to February of 1983, suggesting a qualitative rather than simply incremental adjustment. It also shows that the highest users (families, retired women) made slightly more substantial changes. And it is also noteworthy that the retired women are the one group that moves from relatively high to relatively low consumption between colder and warmer months. This pattern may reflect the location of this group almost exclusively in first-floor apartments, which consistently consume—other relevant factors being equal—more energy in the winter and less in the summer than do the second-story units, but it may also reflect this group's consistent preference for warmth or even a relationship between temperature preference and location.[1]

The information in Figure 1 was derived from the consumption records of 152 of the 240 apartments in the complex. The remaining 88 apartments were occupied by new tenants between January 1983 and February 1984; since we were able to gather data on the social composition of individual apartments at only one point (May 1983), we use here only those apartments whose tenant characteristics were unambiguous. We did, however, re-check the demographic features of the 88 turnover apartments as of February 1984, the end of the period for which we obtained consumption figures; this made possible a comparison of the "movers" and the "stayers" of February 1983 and the "newcomers" and the "stayers" of February 1984. Among these three groups there were no consistent or significant consumption differences—which at least

[1] Although thus far we have been concerned with an interpretation of responses to the billing change, in Table A and Figure A of the appendix we display fourteen months of Multiple Classification Analysis, indicating the relative contributions of various structural and social factors to energy consumption in these units during different seasons. The reader should attend not only to the seasonal changes in energy consumption, but also to the factors that seem to account for it and, indeed, to our ability to account for it at all.

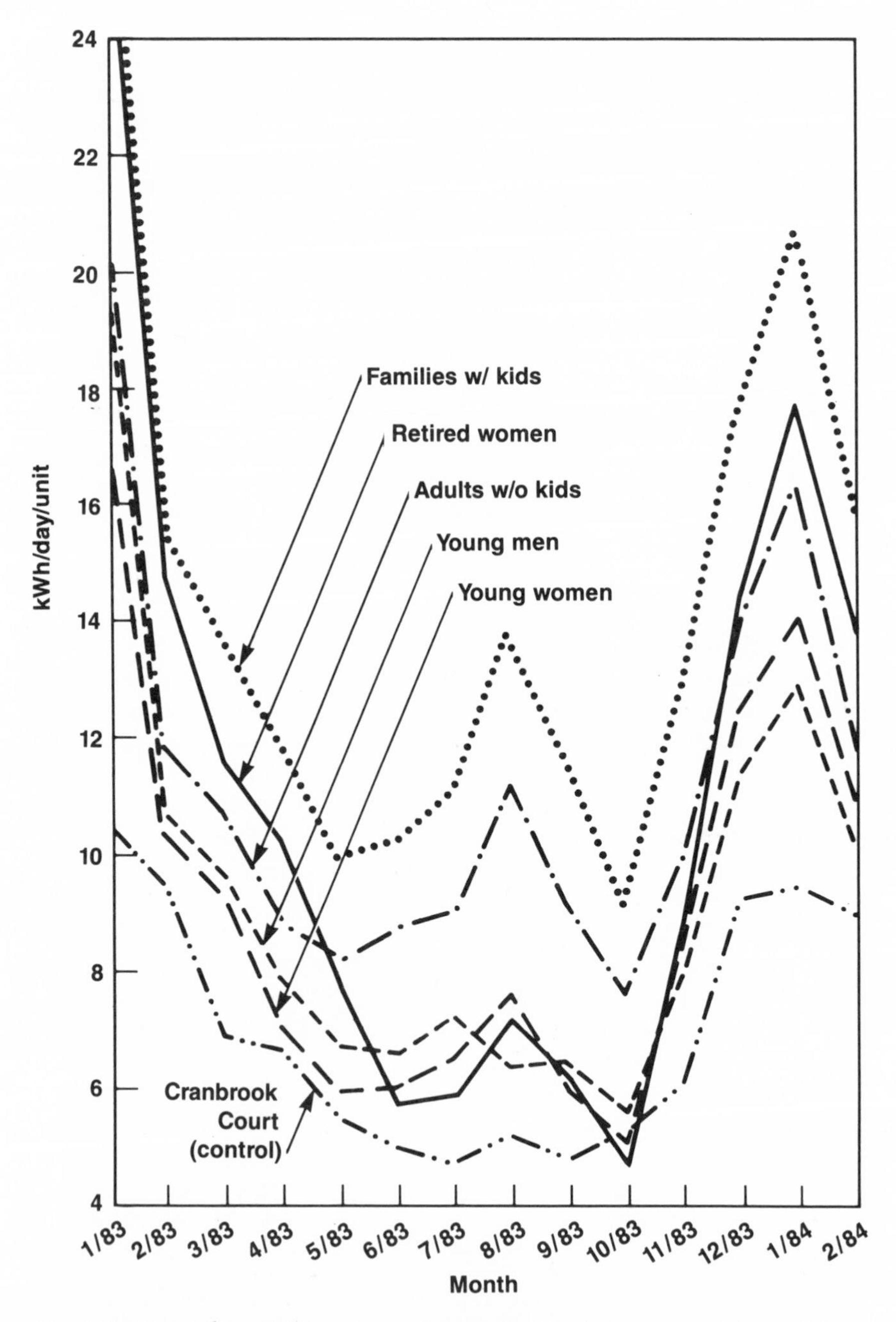


**Fig. 1. Daily apartment kWh consumption for five life-cycle groups, January 1983 to February 1984. Meter conversion occurred in January 1983.**

suggests that the actually going through the conversion (which the newcomers did not do) was not a major factor in determining subsequent consumption.

It will also be noted that the control complex, Cranbrook Court apartments, shows a lower consumption profile than Anderson Place even after the billing change. The reasons for this are probably unexciting: Cranbrook's ceilings are insulated to R-19, Anderson's to R-11; Cranbrook also has a higher proportion of students and somewhat higher rents, so that the apartments there are more likely to be vacated in the summer.

The different types of information presented in Table 1 and in Figure 1 ideally would be combined, so that we could compare consumption for the same 1982 and 1983 months for each life-cycle group, but of course we have no individual apartment data for 1982. The best we could do here was to look at the consumption patterns of the individually-metered *buildings* (each containing several apartments) in the complex, note the change from 1982 to 1983 for each building, and see if we could correlate these change figures with our 1983 data on the demographic characteristics of each building. The results of this procedure were not especially helpful, however. The buildings showed considerable variation in the amount of change in consumption from 1982 to 1983, but a statistical regression analysis of this variation using our demographic data could not explain it.

At the time of the billing change the apartment complex owners rolled back the combined rent-utility payment $30 for each apartment, and the change in consumption patterns meant that most of the tenants lost little time in experiencing a net reduction in their own rent-plus-utility payments. Table 2 indicates the percentage of each group (of "stayers" only) that was saving between $15 and $30 in each of the first fourteen months following the conversion. The differences among the groups indicate the extent to which the tenants of the larger apartments (and, to some degree, the retired women) were being subsidized during the master metering period by other tenants. But given the magnitude of the consumption drop, even during a relatively cold period—the first months of 1983—it seems that consumption is quite elastic for all groups; perhaps consumption was redundant in the period before the change.[2] We were concerned, in the course of this research, that

[2] Much pre-change consumption may have had very low marginal utility. Failure to set back thermostats at night, for example, may consume a good deal of energy in exchange for little comfort.

**Table 2. Percentage of Life-Cycle Groups Saving $15-$30 of the Rent Rollback For Individual Months: January 1983 to February 1984**

| | Life-Cycle Groups | | | | |
|---|---|---|---|---|---|
| | Young Men (N=42) | Young Women (N=23) | Retired Women (N=24) | Two or More Adults W/Kids (N=29) | Families W/Kids (N=34) |
| Jan 83 | 17 | 17 | 4 | 10 | 3 |
| Feb 83 | 52 | 61 | 29 | 45 | 18 |
| Mar 83 | 62 | 61 | 42 | 55 | 24 |
| Apr 83 | 79 | 78 | 50 | 76 | 38 |
| May 83 | 88 | 74 | 67 | 72 | 41 |
| Jun 83 | 90 | 83 | 92 | 72 | 44 |
| Jul 83 | 79 | 78 | 79 | 62 | 41 |
| Aug 83 | 71 | 78 | 71 | 45 | 21 |
| Sep 83 | 83 | 83 | 83 | 59 | 44 |
| Oct 83 | 95 | 91 | 100 | 86 | 56 |
| Nov 83 | 71 | 70 | 54 | 59 | 29 |
| Dec 83 | 45 | 44 | 29 | 31 | 12 |
| Jan 84 | 38 | 26 | 21 | 17 | 9 |
| Feb 84 | 57 | 48 | 29 | 38 | 18 |

the economic rewards of tenant payment to those who consumed the least energy and its value in discouraging profligate consumption might eliminate *de facto* subsidies that seem in some cases to be reasonable–e.g., those to the elderly. But though the elderly did lose their subsidy they also tended in time, as did all the other groups, to make money on the rollback. The fact that tenants' utility costs after the change fell considerably below the level of the rent rollback also might suggest that, from a psychological viewpoint, the opportunity to "make money" through reduced consumption was of more salience than the necessity of "cutting costs" to most of the residents of this apartment complex.

That the change in consumption *persists* is shown by Table 1 and also by the expenditure per group in 1983 as implicitly compared to the same months in 1982 (see Table 2). We expected that the change would be higher in the summer, assuming that coolth is somewhat more expendable to people than warmth–that the association between cold and heating is tighter than the association between heat and cooling–and that the indoor-outdoor temperature differential is usually greater in winter than in summer. Again, however, Davis has a temperate climate, and the California central valley is known for its hot summers, not its cold winters, so the usual relationships may not pertain in this setting. It may also be the case that the billing change, coming as it did in the winter, established in this complex a particularly low winter standard.

## THE INTERVIEWS

In an effort to understand this change in consumption and the variations, both attitudinal and behavioral, in responses to the billing change across our five groups, we interviewed 56 apartment residents, a saturation sample of four buildings. Table 3 offers a compendium of responses to six of the items in the interview. The reliability of the results of these interviews is limited, given the small N's for each group. The results may prove useful, however, if they are compared with research on similar groups in other settings.

Some aspects of the interview responses contrasted with our initial, informal impressions (one research assistant was living at Anderson Place at the time of the billing change) of how tenants would react to the new billing arrangement. In January 1983 there seemed to be widespread anxiety regarding the likely increase in living costs soon to be thrust upon the residents, and this anxiety may account for the especially substantial drop between January and February. By the time of the interviews, in May, there were by contrast relatively few negative views of the new procedures, as Table 3 shows. At the same time, however, many tenants said that a result of the change would be reduced overall comfort in the complex. When we asked about the beneficiaries of the change, most respondents pointed to the local utility company or the apartment complex management, or saw the change as an effort to "control the behavior of the tenants." There was almost no evidence that tenants saw the change primarily as an effort to free the frugal majority from the burdens of the self-indulgent few. Of course interviews may be relatively insensitive to this kind of sentiment, and the culture at large discourages its public expression; but we expected more signs of it than we obtained.

## DISCUSSION

These findings suggest that the mode of payment alone—not simply the affordability of energy, the physical characteristics of dwellings, or the users' wants or needs—can strongly influence levels of energy consumption. We might say that master metering encourages

**Table 3. Percentages of Life-Cycle Groups Reporting Selected Attitudes or Experiences After Billing Change**

| | Life-Cycle Group | | | | |
|---|---|---|---|---|---|
| Comments | Young Men | Young Women | Retired Women | Adults W/O Kids | Families W/Kids |
| Had negative reactions to change | 0 | 0 | 25 | 29 | 12 |
| Saw the change as reducing tenant comfort | 50 | 56 | 57 | 62 | 47 |
| Saw the billing change as | | | | | |
| - an aid to owners and the utility company | 40 | 18 | 22 | 29 | 29 |
| - an aid to the tenants | 10 | 9 | 11 | 0 | 6 |
| - means of controlling tenants | 10 | 46 | 67 | 43 | 35 |
| Showed no evidence of behavioral change due to billing change | 44 | 22 | 38 | 25 | 13 |
| Cannot further reduce electricity consumption | 80 | 46 | 56 | 50 | 39 |
| Could respond to "Electricity Shortage" with reduced use | 90 | 91 | 89 | 75 | 82 |
| | N=(10) | (11) | (9) | (8) | (18) |

waste or redundancy in energy use, whether that wastefulness be inadvertent or intentional. I speculate, however, that both waste and frugality become meaningful to consumers primarily in a kind of game: under master metering, the object is to maximize the amount of energy consumed for the number of dollars committed in advance; under individual metering, the object is to minimize the number of dollars spent to meet already established standards of comfort or consumption. These standards, moreover, are flexible—matters of social definition and control. As an illustration of this concept, the last two items in Table 3 suggest together that avowed consumption requirements are at least in part a function of how we asked questions about them. People are inclined to believe that their consumption levels are "minimal," but virtually everyone agrees that these "needs" could be cut further if there were a (in this case, vaguely defined) "shortage of electricity." Thus it seems important that we not favor individual metering to the extent of assuming that it eliminates the inflation of consumption; energy requirements are, in this sense, established by considerations of appropriateness and social normality.

A broad sociological assumption here is that human monitoring devices—neighbors and sociologists, for example—are important in regulating consumption levels. We were concerned about this problem in interpreting our interviews, since interviews monitor what is *said* about behavior, and one might expect a considerable discrepancy between what is said and what is done. We assumed that people would paint an idealized picture, and tested this assumption by asking our interviewees how their energy consumption levels compared with those of their neighbors and comparing this report with actual energy consumption levels. The results of this effort were ambiguous, but we realized, in any case, that a tenant could always choose a neighbor whose consumption was particularly high and count himself, by comparison, frugal and responsible. It is nearly a sociological truism that people tend to maintain an idealized picture of their own actions. Interviewees can give even poor performances felicitous interpretations as "temporary," required by "special circumstances," or warranted by "exceptional events" (Hackett et al. 1984). These interpretations may explain the discrepancy in responses to two of our interviewing questions: we asked people how many loads of laundry they "usually" do in a week, and then asked them how many they had done "last week." For only 13 of the 35 respondents who gave "modal" responses (one, two, or three loads "usually") did the two answers match.

Consumption is governed largely by considerations of reputation. A reputation may, of course, adhere to a dwelling as the result of inquiry into its performance, but that performance becomes important mainly as it bears on the performance and reputation of the dweller. How people actually manage their dwellings as well as their reputations ought then to be a central focus of energy consumption studies (Kempton 1982; Diamond 1983). How dwellings are managed will be a function of the nature of their billing systems and perhaps especially of the meanings attached to the appliances whose use these systems record. For this work we need a cross-cultural or anthropological perspective; from preliminary interviews for an energy study in a housing complex with a large number of "third world" students, for example, we have already been firmly reminded of the important place that large electrical appliances have as symbols of modernity. These are matters of social definition and social control, and the importance of social control in energy use implies that consumption levels are indeed elastic, probably radically so.

Finally, the kind of research reported here might have some value for relating energy use to demographic trends. In particular, the study of single-resident dwellings or of non-familial housing seems of special importance because of the current, rapidly increasing demand for this kind of living situation (Simms 1983). While single-resident households (and related appliances such as non-family automobiles) promote the values of compactness and efficiency (relative, at least, to the ancestral models), they clearly increase resource consumption *per individual.* They may also be, since less social, more difficult to monitor. One of the major issues of domestic energy consumption in the future, then, may be the question of how to encourage the sharing of resources and otherwise to socially control resource consumption in non-familial communities.

## Acknowledgements

*Two exceptional University of California, Davis undergraduate sociology majors contributed to this paper: Michael Harrington originally proposed the project and made crucial contributions at every stage; Alice Kwong joined the project late, but helped in shaping all aspects of the final product. Financial support for this research came from the University of California Universitywide Energy Research Group and is gratefully acknowledged.*

## References

Diamond, R.C. "Energy and Housing for the Elderly: Preliminary Observations." Paper presented at the Working Conference on Families and Energy, Michigan State University, October 1983.

Hackett, B., P. Craig, J. Cramer, T. Dietz, D. Kowalczyk, M. Levine and E. Vine, "Comparing the Methodologies of Research on Household Energy Consumption." Washington, DC: American Council for an Energy-Efficient Economy, 1984.

Kempton, W. "Do Consumers Know 'What Works' in Energy Conservation?" In *What Works: Documenting Energy Consumption in Buildings*, ed. J. Harris and C. Blumstein. Washington, DC: American Council for an Energy-Efficient Economy, 1984.

McClelland, L. "Short-term Effects of Tenant Payment of Utilities on Energy Use in Multifamily Housing." Paper read at the ACEEE Second Summer Study on Energy-Efficiency in Buildings, University of California, Santa Cruz, August 1982.

Nelson, S. "Energy Savings Attributable to Switching from Master to Individual Metering of Electricity." Argonne, IL: Argonne National Laboratory, 1981.

Simms, M.C. "Women and Housing: The Impact of Government Housing Policy." In *Families, Politics and Public Policy*, ed. I. Diamond. New York: Longman, 1983.

## Appendix

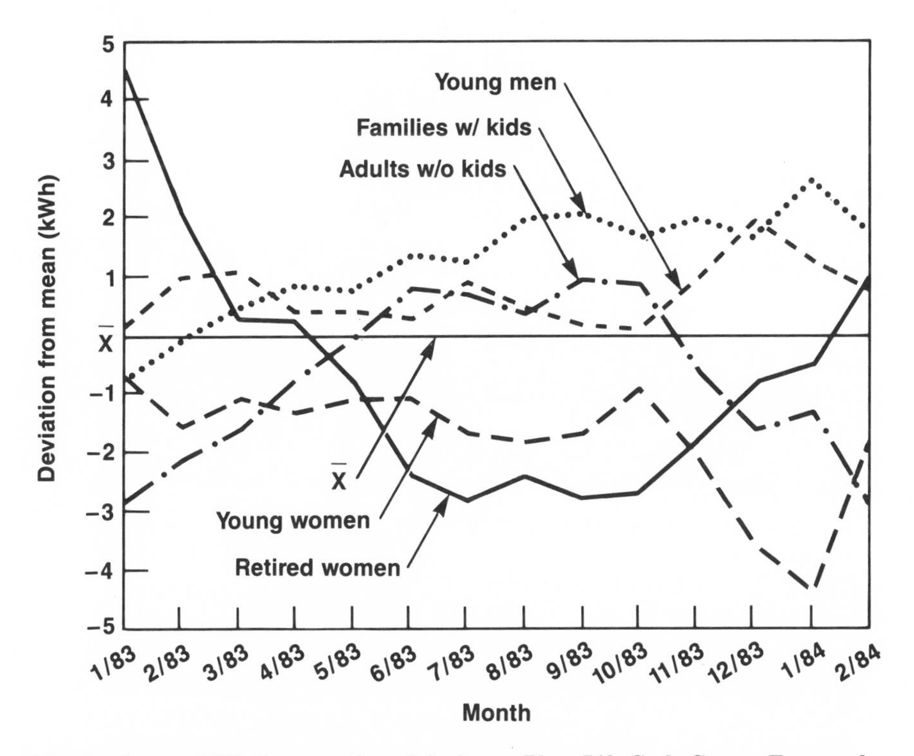


**Fig. A. Average kWh Consumption of Anderson Place Life Cycle Groups Expressed as Adjusted Standard Deviations from Monthly Total Population Means (SPSS Multiple Classification Analysis).**

**Table A. Multiple Classification Analysis of 1983 Anderson Place Apartment kWh Consumption Using Life-Cycle, Income, and Location Variables**

| | Months | | | | | | | | | | | | | |
|---|---|---|---|---|---|---|---|---|---|---|---|---|---|---|
| | Jan 83 | Feb 83 | Mar 83 | Apr 83 | May 83 | Jun 83 | Jul 83 | Aug 83 | Sep 83 | Oct 83 | Nov 83 | Dec 83 | Jan 84 | Feb 84 |
| GRAND MEAN | 21.86 | 12.51 | 11.02 | 9.17 | 7.71 | 7.61 | 8.01 | 9.64 | 8.00 | 6.50 | 9.70 | 14.01 | 16.38 | 12.54 |
| **LIFE CYCLE** | | | | | | | | | | | | | | |
| Young Men (41) | 0.23 | 1.02 | 1.11 | 0.42 | 0.37 | 0.30 | 0.94 | 0.52 | 0.16 | 0.13 | 0.99 | 2.01 | 1.34 | 0.98 |
| Young Women (21) | -0.73 | -1.55 | -1.19 | -1.38 | -1.08 | -1.08 | -1.68 | -1.85 | -1.71 | -0.91 | -2.03 | -3.55 | -4.40 | -1.85 |
| Retired Women (24) | 4.56 | 2.13 | 0.29 | 0.31 | -0.79 | -2.38 | -2.75 | -2.41 | -2.79 | -2.72 | -1.85 | -0.80 | -0.54 | 0.96 |
| Adults W/O Kids (28) | -2.85 | -2.06 | -1.56 | -0.82 | 0.02 | 0.84 | 0.73 | 0.43 | 1.00 | 0.92 | -0.62 | -1.56 | -1.32 | -2.86 |
| Families W/Kids (32) | -0.75 | -0.08 | 0.51 | 0.85 | 0.82 | 1.37 | 1.31 | 1.98 | 2.14 | 1.66 | 1.99 | 1.72 | 2.73 | 1.75 |
| **INCOME** | | | | | | | | | | | | | | |
| $0 – $4,999 (62) | -1.21 | -0.57 | -0.20 | -0.32 | -0.31 | -0.48 | -0.51 | -0.77 | -0.66 | -0.55 | -0.13 | -0.20 | -0.96 | -0.64 |
| $5,000 – $9,999 (66) | -0.69 | -0.41 | -0.47 | -0.19 | 0.04 | 0.24 | 0.04 | 0.19 | 0.23 | 0.44 | -0.07 | -0.23 | 0.46 | 0.09 |
| $10,000 or more (18) | 6.72 | 3.48 | 2.43 | 1.79 | 0.93 | 0.79 | 1.59 | 1.97 | 1.45 | 0.30 | 0.72 | 1.52 | 1.64 | 1.89 |
| **LOCATION** | | | | | | | | | | | | | | |
| Upstairs (70) | -0.88 | -1.16 | -0.59 | -0.85 | -0.58 | 0.02 | 0.11 | 0.22 | -0.39 | -0.58 | -0.20 | -0.40 | -0.97 | -0.96 |
| Downstairs (76) | 0.81 | 1.06 | 0.55 | 0.78 | 0.53 | -0.02 | -0.10 | -0.21 | 0.36 | 0.53 | 0.18 | 0.37 | 0.89 | 0.88 |
| MULTIPLE $R^2$ | 18% | 21% | 15% | 20% | 25% | 28% | 31% | 31% | 32% | 35% | 16% | 12% | 15% | 15% |

# Statistical Analysis of Lifestyle Factors in Heating Energy Use of New and Weatherized Minnesota Houses

*Mary H. Fagerson*
*Minnesota Department of Energy and Economic Development*

## INTRODUCTION

Heating energy use is known to vary significantly even among identical houses, and the lifestyle of the occupant is commonly assumed to be the major cause of such variation. This study analyzes the effects of certain characteristics of two groups of houses and their occupants on the heating energy use of these houses. The house apparently can affect lifestyle; and both the house design and the lifestyle of its occupants can influence the heating energy use of the house.

## DESCRIPTION OF THE TWO SAMPLES

The data for this study are taken from two sets of houses. The first sample is a group of 46 houses from the Energy-Efficient Housing Demonstration Program (EEHDP) of the Minnesota Housing Finance Agency (MHFA). They are located primarily within 100 miles of Minneapolis/St. Paul, but a few are in or near Moorhead, Minnesota.

The EEHDP is described in the preliminary analysis by Hutchinson et al. (1982). One hundred forty-four homes were built under

*Energy Efficiency: Perspectives on Individual Behavior*

this program. Data available on these houses include utility data, blower door test results, house plans, and homeowner questionnaires. Energy-use data for the 1981-82 heating season were available on 98 of these homes. However, because of missing blower door tests and homeowner questionnaires, the database used for this analysis was limited to 46 homes. An analysis by Fagerson and Lancaster (1983) divides this sample into two subsamples: those with a large area of south glass in proportion to the floor area of the home and those with a small area of south glass.[1]

The other group is a sample of 23 houses that were weatherized under the Weatherization Program. The weatherized sample is all within the greater metropolitan area of the Twin Cities. These houses are each of a different design and range in age from 10 to over 100 years. The types of work done on these homes included the addition of wall, ceiling, and rim joint insulation, and in a few cases exterior basement wall insulation. Broken windows were replaced and storm windows and storm doors added. Clock thermostats were added in some cases. Almost all homes received caulking and weatherstripping. In no cases were these homes brought to a level that exceeds the requirements of the 1978 Minnesota Energy Code for new construction. Some houses are still without wall insulation. An analysis of the energy use and the various weatherization strategies was carried out by June Wheeler and Peter Hertzog of Bakke, Kopp, Ballow & McFarlin, Inc. (1983) under contract with the Department of Economic Security. Further data on 35 of these houses were collected by the author of this paper, using blower door tests, infra-red scans, house measurements, and homeowner questionnaires.[2]

---

[1] The analysis by Fagerson and Lancaster developed a multiple regression model using the actual heat load as the dependent variable and calculated values of heat loss and gain as the predictor variables. The hypothesis was that the coefficient of the heat loss variables would be +1 and that of the heat gain variables would be -1. The sample of 46 houses was divided into two subsamples: houses with high south-glass-to-floor-area ratios and those with low south-glass-to-floor-area ratios. The results indicated that the model was better at predicting the energy use of the low south glass sample better than the high south glass sample.

[2] A blower door test provides a measurement of the leakiness of a house. It consists of pressurizing and/or depressurizing a house by means of a fan sealed in a doorway, and measuring the air flow of the fan required to create various pressure differentials across the envelope of the house. Infra-red scans are conducted by using infra-red thermography to observe warm or cold surfaces within the house. Use of the blower door to pressurize or depressurize the house during an infra-red inspection accentuates points of air leakage. Measurements of walls and windows were taken using a tape measure. The homeowner questionnaire was filled in by the author during on-site interviews with house occupants.

## DATA ANALYSIS

A data file of up to 60 variables describing each house in the two samples, its climate, and its occupants was created. Table 1 lists and defines the primary variables used in this analysis. Analysis of the two samples was carried out using MULTREG, a statistical package developed at the University of Minnesota. Analysis of the

**Table 1. VARIABLE LIST – The column on the left indicates the house sample for which the variable was available as calculated from available data.** (E stands for homes in the Energy Efficient Housing Demonstration Program, and W stands for homes retrofitted under the Weatherization Program)

| Available for House Sample | Variable |
|---|---|
| E, W | Calculated below grade heat loss (MBTU/yr)* |
| E, W | Calculated total above grade heat loss (MBTU/yr)* |
| E | Calculated heat loss from above grade walls only (MBTU/yr) |
| E | Calculated heat loss from ceiling and roof only (MBTU/yr) |
| E | Calculated heat loss from non-south windows only (MBTU/yr)* |
| E, W | Calculated net passive solar gain (MBTU/yr)* |
| E, W | Calculated internal gains from people (MBTU/yr)* |
| E, W | Calculated internal gains from appliances (MBTU/yr)* |
| E, W | Air flow at 50 Pascal pressure difference (CFH) |
| E, W | Actual heat load (the product of heating fuel used minus the baseload and AFUE rating) (MBTU/yr)* |
| E | Attached unit (these units were townhouse units attached in groups of two to four units) |
| E, W | Family size (the number of people living in the house) |
| E, W | People-hours (the sum of the average number of hours each person occupying the house was home each day) |
| E, W | Thermostat setting (the average reported 24-hour thermostat setting in °F) |
| E | Night insulation use (the number of hours per day insulation covered house windows) |
| E | Hot water usage (gallons of hot water used per week calculated from information on cooking, clothes washing, and bathing). |
| E | Variation in energy use (difference between the average heat load for a design and the heat load of a specific house of that design) (MBTU/yr) |
| W | Total surface area above grade (sq. ft.) |
| E, W | Above grade wall and ceiling heat loss (ΣUA) |
| E, W | Below grade wall heat loss (ΣUA) |

* See Fagerson and Lancaster (1983) for a description of methods used to calculate these values.

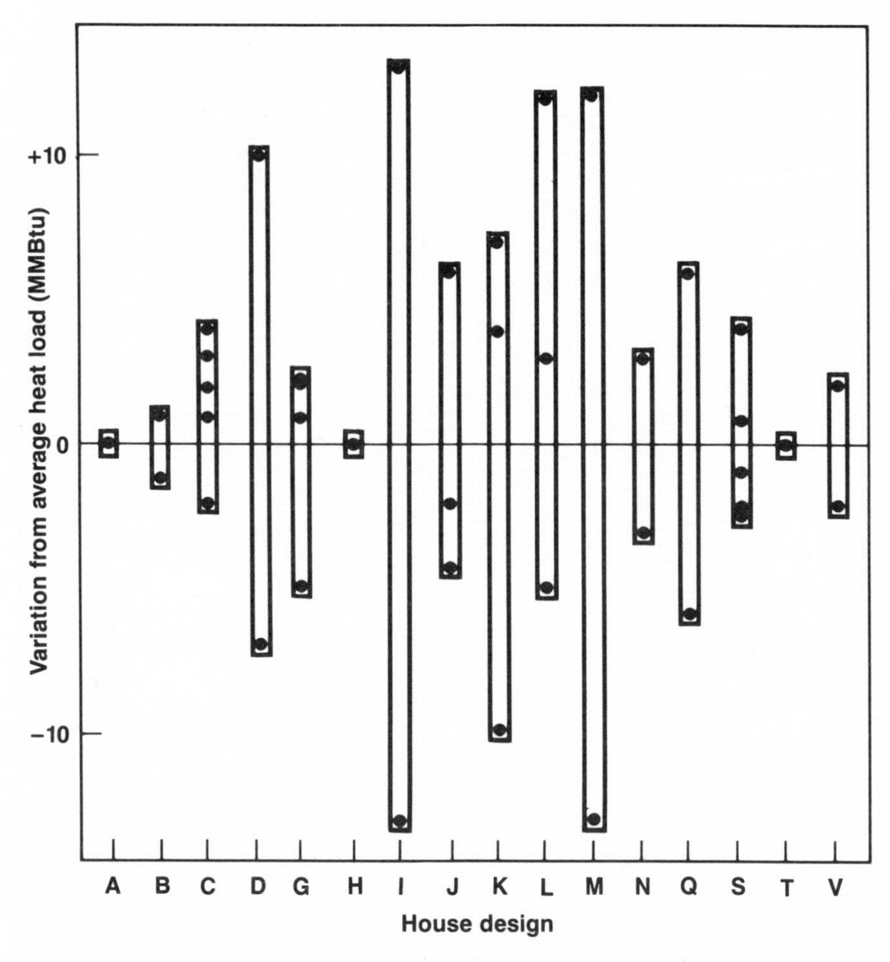


**Fig. 1. Variation from average heat load for each design in the EEHDP sample.**

EEHDP houses included correlations and multiple regressions of the actual heat load and the variation of the heat load from the average of identical houses with the characteristics of the house and its occupants. The second part of the analysis was a comparison of the relationships identified in the energy-efficient houses with the relationships in the weatherized house group.

### Variation of Heating Energy within a Design Group

Figure 1 is a plot of the variation in energy use for each design in the EEHDP sample. The mean heating load for each design is located on the horizontal line. The dots within the bar indicate the

**Table 2. Correlations with variations from average heating for each design and significance levels. (The sample size is 46.)**

| | r | Significance |
|---|---|---|
| Thermostat Setting | 0.44 | 0.01 |
| Number of People | -0.16 | 0.10 |
| People-Hours | -0.15 | 0.10 |
| Attached Units | -0.03 | 0.10 |
| Air Change Rate | 0.06 | >0.10 |

variation of individual houses from the average of that design. Taller bars indicate designs with larger variations.

The correlations (Table 2) indicate that the most significant characteristics influencing variation within a design group are the thermostat setting, the number of residents, the number of hours the house was occupied, and whether or not the unit was attached to another unit. A multiple regression of the variation in energy use on these variables (Figure 2) explains only 32% of variation in the dependent variable. If the EEHDP sample is divided into the two

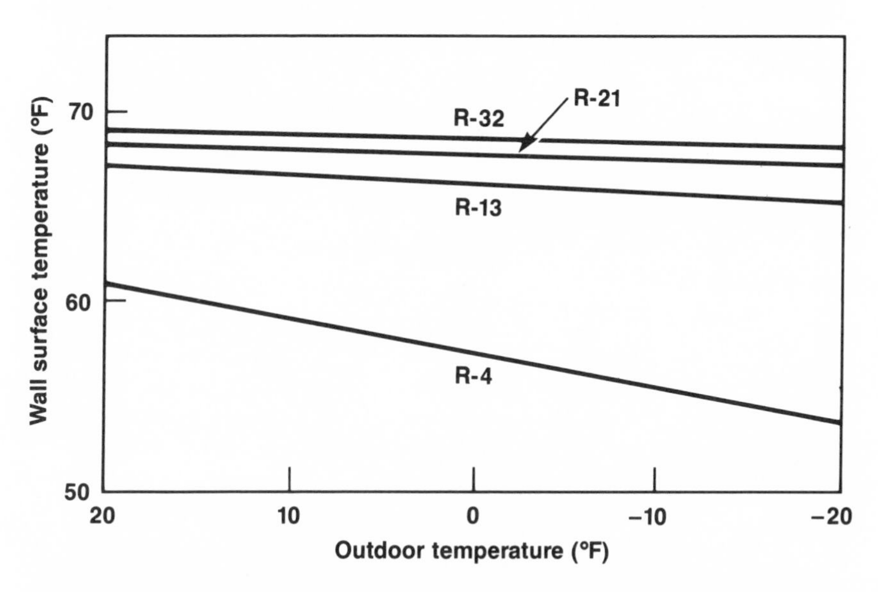


**Fig. 2. Calculated interior wall surface temperatures at various outdoor temperatures when the indoor air temperature is 70° F for several levels of wall insulation.**

**Table 3. Multiple regression of variation in energy use on thermostat settings, people-hours, and attached units in EEHDP house sample.**

**Full Sample**

| | VARIABLE | COEFF. | STD ERROR | T VALUE |
|---|---|---|---|---|
| Constant | B0 | -53.87854 | 14.22335 | -3.79 |
| People-hours | V19 | -0.8906457E-01 | 0.3518431E-01 | -2.53 |
| Thermostat | V22 | 0.9001180 | 0.2246009 | 4.01 |
| Attached | V25 | -2.738268 | 1.712375 | -1.60 |

degree of freedom= 38
residual mean square= 25.06756
root mean square= 3.006752
r-squared= 0.3217

**High South Glass Sample**

| | VARIABLE | COEFF. | STD ERROR | T VALUE |
|---|---|---|---|---|
| Constant | B0 | -72.18682 | 20.53975 | -3.51 |
| People-hours | V19 | -0.6475174E-01 | 0.8275856E-01 | -0.78 |
| Thermostat | V22 | 1.184043 | 0.3433717 | 3.45 |
| Attached | V25 | -4.222806 | 2.760948 | -1.53 |

degree of freedom= 19
residual mean square= 32.12129
root mean square= 3.667564
r-squared= 0.4085

**Low South Glass Sample**

| | VARIABLE | COEFF. | STD ERROR | T VALUE |
|---|---|---|---|---|
| Constant | B0 | -13.72912 | 20.29177 | -0.68 |
| People-hours | V19 | -0.9902582E-01 | 0.3163538E-01 | -3.13 |
| Thermostat | V22 | 0.2896741 | 0.3077650 | 0.94 |
| Attached | V25 | -1.917070 | 1.998851 | -0.96 |

degree of freedom= 15
residual mean square= 14.68877
root mean square= 3.832593
r-squared= 0.4055

**Table 4. Analysis of variance of regression of variation in energy use on thermostat setting, people-hours, and attached units in EEHDP sample.**

**Full Sample**

| | SOURCE | INDIVIDUAL DF | INDIVIDUAL SS | INDIVIDUAL MS | CUMULATIVE DF | CUMULATIVE SS | CUMULATIVE MS | $R^2$ |
|---|---|---|---|---|---|---|---|---|
| | MEAN | 1 | 2.3810 | 2.3810 | | | | |
| Thermostat | V22 | 1 | 278.84 | 278.84 | 1 | 278.84 | 278.84 | 0.19855 |
| Attached | V25 | 1 | 12.375 | 12.375 | 2 | 291.21 | 143.61 | 0.20736 |
| People-hours | V19 | 1 | 160.63 | 160.63 | 3 | 451.84 | 150.61 | 0.32173 |
| | RESIDUAL | 38 | 952.57 | 25.068 | 41 | 1404.4 | 34.234 | |

**High South Glass Sample**

| | SOURCE | INDIVIDUAL DF | INDIVIDUAL SS | INDIVIDUAL MS | CUMULATIVE DF | CUMULATIVE SS | CUMULATIVE MS | $R^2$ |
|---|---|---|---|---|---|---|---|---|
| | MEAN | 1 | 4.3565 | 4.3565 | | | | |
| Thermostat | V22 | 1 | 346.02 | 346.02 | 1 | 346.02 | 346.02 | 0.33535 |
| Attached | V25 | 1 | 55.812 | 55.812 | 2 | 401.83 | 200.91 | 0.38945 |
| People-hours | V19 | 1 | 19.664 | 19.664 | 3 | 421.49 | 140.50 | 0.40850 |
| | RESIDUAL | 19 | 610.30 | 32.121 | 22 | 1031.8 | 46.900 | |

**Low South Glass Sample**

| | SOURCE | INDIVIDUAL DF | INDIVIDUAL SS | INDIVIDUAL MS | CUMULATIVE DF | CUMULATIVE SS | CUMULATIVE MS | $R^2$ |
|---|---|---|---|---|---|---|---|---|
| | MEAN | 1 | 0.52632E-05 | 0.52632E-05 | | | | |
| People-hours | V19 | 1 | 123.05 | 123.05 | 1 | 123.05 | 123.05 | 0.33200 |
| Attached | V25 | 1 | 14.242 | 14.242 | 2 | 137.29 | 68.646 | 0.37042 |
| Thermostat | V22 | 1 | 13.013 | 13.013 | 3 | 150.31 | 50.102 | 0.40553 |
| | RESIDUAL | 15 | 220.33 | 14.689 | 18 | 370.64 | 20.591 | |

subsamples used by Fagerson and Lancaster (1983), high south glass houses and low south glass houses, the relationships change somewhat. A multiple regression (Table 3) of the variation in energy use on thermostat setting, occupied people-hours, and attached units indicates that in the houses with high south-glass-to-floor-area ratios thermostat setting and attached units are the significant factors. In the low south-glass house samples, thermostat setting alone explains 34% of the variation and attached units explains 5%. In the low south glass sample, people-hours explains 33% of the variation (see Table 4).

Air change rate is not a significant factor in explaining the variation among identical houses when thermostat setting or people-hours are taken into account. The other known occupant characteristics are also insignificant in this model. The characteristics explored are family size, appliance heat gain, gallons of hot water used per week, and night insulation use. There are other variations in occupant behavior that do not occur frequently enough to be included in this analysis. These variations include responses to condensation on windows or to overheating, which occurred in a few of the houses with many windows.

### Comparison of the Energy-Efficient Houses with the Weatherized Houses

A comparison of the relationship between the air change rate and heat load in the weatherized and the energy-efficient houses is shown in Table 5. It is also interesting to look at the relationships of thermostat settings to air change rates and heat loss levels (U-value) both above and below grade. Thermostat settings in the weatherized houses appear to be closely linked to the air change rate ($r = 0.60$) and to a lesser degree to the heat loss (U-value) of walls ($r = 0.49$ and 0.45). The air change rate (ach) in the weatherized houses ranges from 3.6 to 11.8 at 50 Pascals and averages 8.8, while that of the energy-efficient houses ranges from 2.2 to 8.9 and averages 4.4 (see Table 6). The above-grade wall insulation levels range from R-4 to R-11 in the weatherized houses and from R-20 to R-33 in the energy-efficient houses.

## DISCUSSION AND ANALYSIS

This analysis focuses on two areas: the variation in energy use among identical houses and the apparent influence of the building's design on the lifestyle of its occupants.

**Table 5. Correlations in the weatherized and energy-efficient house samples and significance levels.** (Sample sizes are 25 and 46, respectively.)

| | Weatherized | | Energy Efficient | |
|---|---|---|---|---|
| | r | Significance | r | Significance |
| Heating with air flow at 50 Pa | 0.70 | 0.001 | 0.55 | 0.001 |
| Thermostat setting with air flow | 0.60 | 0.01 | 0.20 | >0.10 |
| Thermostat setting with above grade heat loss (ΣUA) | 0.49 | 0.02 | 0.06 | >0.10 |
| Thermostat setting with below grade wall heat loss (ΣUA) | 0.45 | 0.05 | 0.17 | >0.10 |

**Table 6. Comparison of air change rates and insulation levels in the two samples.**

| | Weatherized | | | Energy Efficient | | |
|---|---|---|---|---|---|---|
| | Min. | Ave. | Max. | Min. | Ave. | Max. |
| Air change rate (ach) | 3.6 | 8.8 | 11.8 | 2.2 | 4.4 | 8.9 |
| Above grade wall R-value | 4 | | 13 | 20 | | 33 |
| Below grade wall R-value | 2 | | 11 | 10 | | 29 |

### Variation in Energy Use of Identical Energy-Efficient Houses

The variation in energy use among identical energy-efficient houses from the average for that design ranges from 1% to 40% of the average. At least 32% of this variation may be explained by some of the lifestyle or design factors examined in this analysis. Construction quality (indicated by blower door test results) does not appear to be a significant factor in explaining the variation.

The significant lifestyle and design factors identified in the analysis of the EEHDP houses affect some houses more than others. Apparently high south glass houses are influenced more by the lifestyle and design factors that affect heat loss (thermostat setting and attached units) while low south glass houses are primarily influenced by occupied people-hours, which affect the internal gains. The significance of people-hours may be a result of body heat given off by the occupants, heat dissipated from showers and other hot water usage, or heat given off by appliances. People-hours are strongly correlated with gallons of hot water used per week and to a lesser degree appliance use ($r = 0.64$ and $0.15$). People-hours are also somewhat correlated to thermostat setting ($r = 0.26$). The significance of these findings is that not only can the occupant behavior (thermostat setting and people-hours) influence the energy use of a house, but the design of the house (attached units and south glass area) can interact with lifestyle to influence the energy use. Sixty-eight percent of the variation in energy use among identical houses in the full sample or 60% in the subsample remains unexplained. The unexplained variation may be due either to building and lifestyle characteristics that are not explored in this analysis or to imprecise measurement of the characteristics that are explored.

Throughout this study, similar units by one builder are referred to as identical, but there are several important differences among these houses. First, there are slight variations in design in one set of townhouse units. There are also differences caused by changes that the occupants of the houses have made. Some of these changes may increase energy use, such as adding a dryer vent or another door to the exterior. Other changes may reduce energy use, such as added insulation, caulking or weatherstripping. A third source of variation is the proper functioning of the mechanical equipment. One homeowner complained of a malfunctioning "automatic switch" on the furnace, others of malfunctioning heat exchangers. None of these differences in nominally identical buildings have been included in the analysis. Some of the unexplored differences in behavior include occupant responses to condensation on windows or to overheating. Another source of variation among identical units may be the behavior of the occupants of the neighboring units in attached townhouses. Spielvogel (1983) has shown that the heating use of an attached unit may be influenced by the temperature levels of the unit or units attached to it. The present analysis did not take into account the temperatures of neighboring units.

There are several possible sources of error in the variables used in this analysis. The calculation of infiltration heat loss is based on blower door tests and may not provide an accurate estimate of infiltration heat loss. The air flow rate at 50 Pascals was divided by 15 to estimate natural infiltration rates. However, this approach assumes that the leakage areas in each house are behaving in a similar manner. The locations of the leakage areas may cause significant differences in the infiltration rates of these houses. One group of houses, not included in the sample discussed in this study, Design E, has fairly high leakage areas (7.55-9.37 ach at 50 Pa) and yet the heating energy used is quite low (1.87-3.34 BTU/sq.ft.-DD). A second reason that the blower test results may not provide accurate estimates of infiltration heat loss is that they provide a measure only of the leakiness of the house and not of the occupant's use of the house (such as frequency of door openings, etc.). One cause of an occupant's opening the doors is to let pets in and out. Only 14 houses in the energy-efficient house sample had pets. The occupants said they let pets in and out from 3 to 7 times per day. However, there was not a significant correlation between the number of times pets were let in and out and the heating load for these 14 houses.

A second source of error may be the data used to estimate lifestyle effects. The occupants of the energy-efficient houses were sent a questionnaire during January of the year for which energy-use data are available. They were asked about thermostat settings and schedules and unheated rooms in the house. Many of these habits are difficult to estimate for the whole season and may be subject to change for various reasons.[3]

Third, the method of analysis makes the assumption that thermostat settings are equivalent to the interior temperatures that drive the heat loss of the house. However, one month of temperature recordings in six of these houses indicate that temperatures are

[3] Data from 7 houses collected by Kempton and Krabacher at the Institute for Family and Child Study in Michigan (this volume) demonstrate how difficult it may be for some occupants to estimate an average thermostat setting for the whole year. Occupants of some houses make many adjustments in the thermostat setting throughout each day. Occupants reported average thermostat settings at least 2°F below actual settings.

not uniformly distributed throughout these houses.[4] Thermostat settings may not be equal to the average indoor temperature.

Even with the error in the temperature variable used in this analysis, this variable is the most significant factor explaining the variation in energy use among identical designs with large south glass areas. A study of the relationships among the stated thermostat settings, the actual thermostat settings, and the actual temperature or distribution of temperatures within houses might be able to explain a larger portion of the variation in energy use among identical houses.

**Building Effects on Lifestyles in Weatherized Houses**

The lifestyle of the occupant may influence the energy use of the house. However, the infiltration rates and insulation levels of a house may also influence the occupant's thermostat setting. In weatherized houses this effect appears to be significant. Apparently in draftier houses or houses with cold wall surfaces, the occupant needs a higher air temperature to be comfortable. Frequently, high air change rates were associated with large attic bypasses. In one case an occupant mentioned how drafty it had been in the dining area next to a built-in buffet. The back of the buffet was connected to the attic through the uninsulated wall cavities. In another house whose air temperature was 75°F and whose walls were uninsulated, the owner complained of always feeling cold.

The comfort level of the house is partially dependent on the mean radiant temperature of surfaces. According to Christopher Alexander (1977), experimentation has established that the most comfortable balance between air temperature and the surrounding surfaces occurs when the mean radiant surface temperature is two degrees higher than the ambient temperature. The temperature of exterior wall surfaces in uninsulated houses is much more dependent on the outside temperature than that of walls insulated to total R-values of R-13 or more (see Figure 2). It would be interesting to monitor the thermostat settings in insulated and uninsulated houses to see if, as the outside temperature decreased, the average indoor thermostat setting increased more in uninsulated houses

---

[4] Temperature recording devices were placed in at least six locations of the EEHDP houses by a University of Minnesota engineering student under the direction of the author. The data were to be used to identify appropriate locations for recording devices in the current research on the EEHDP houses. The data tapes were analyzed by Peter Cleary at Lawrence Berkeley Laboratory.

than in insulated houses.

The correlation between thermostat setting and airflow or wall U-value is not significant in the energy-efficient houses. In existing houses, it should be possible to bring air change rates and wall U-values to the level at which occupants can be comfortable without high thermostat settings. If bypasses located near activity centers of a house are the cause of higher thermostat settings, it should be possible to repair these enough to allow lower thermostat settings. If surface temperatures of uninsulated walls are the cause of high settings, adding R-11 insulation should make these houses less susceptible to changes in exterior temperatures than are attached houses with low south glass areas.

## CONCLUSION

Heating energy use can vary significantly even among presumably identical energy-efficient houses. This study raises four issues in understanding the causes of variations in energy use. First, a large portion of the variation among identical houses is unexplained by the variables considered in this analysis. This result may be partly due to the fact that truly identical houses may not exist. Since houses are built by individuals and inhabited by individuals, there are many variations in the house and the way it is used. The many possible combinations of these variations make the existence of truly identical houses unlikely.

Second, although thermostat setting is an imprecisely measured variable and not an accurate representation of the actual temperature in a house, it still can explain one fifth of the variation in the energy use of identical houses.

Third, the variation in energy use among nominally identical houses may be due not only to variations in lifestyle but also to certain characteristics of a house design that make it more responsive to variation in lifestyle. Single-family detached houses and houses with large glass areas may be more susceptible to variation in energy use because of variation in, for example, thermostat setting.

Fourth, the design of the house can determine which lifestyle factors will have the most significant impact on the energy use of the house. Houses whose energy use is dominated by heat loss will be most strongly affected by the thermostat setting, while houses whose energy use is dominated by heat gains will be most strongly affected by the number of hours the house is occupied by each family member.

Finally, occupant lifestyle may be somewhat dependent on building design and construction. In particular, thermostat settings may be influenced by the location of bypasses or by the lack of insulation in walls. All new houses must be insulated to comply with the Minnesota Energy Code. However, bypasses may occur in new houses as well as old houses if the designer and builder are not aware of the significance of bypasses into unheated spaces. The education of owners, designers, and builders of houses about both the direct effects of insulation and bypasses on heat loss of the house and the indirect effect of these factors through thermostat settings would be valuable. Also, further study of the interrelationship of comfort levels, lifestyle, and building characteristics may be valuable to our understanding of variations in the energy use of houses.

**References**

Alexander, C. *A Pattern Language.* New York: Oxford University Press, 1977.

Fagerson, M. "Field Study of Thirty-Five Houses in the Minnesota Weatherization Assistance Program." Unpublished draft, October 1983.

Fagerson, M. and R. Lancaster. "A Statistical Analysis of Passive Solar Superinsulated Homes in Minnesota." In *Proceedings of the American Solar Energy Society.* Minneapolis: American Solar Energy Society, June 1983.

Hutchinson, M. G. Nelson and M. Fagerson. "Measured Thermal Performance and the Cost of Conservation for a Group of Energy-Efficient Minnesota Homes." Paper presented at the Second ACEEE Summer Study on Energy Efficiency in Buildings, Santa Cruz, CA, August 1982.

Spielvogel, L. "Garden Apartment Energy Use." *Heating/Piping/Air Conditioning*, September 1983.

Wheeler, J. and P. Hertzog. "A Study of the Effectiveness of the Weatherization Program in Minnesota." Minnesota Department of Economic Security, January 1983.

Young, H.D. *Statistical Treatment of Experimental Data.* New York: McGraw-Hill Paperbacks, 1962.

Weisberg, S. "MULTREG Users Manual." University of Minnesota School of Statistics, Technical Report Number 298, Summer 1981.

# INDEX